THE PENDULUM AND THE TOXIC CLOUD

also by Thomas Whiteside

THE RELAXED SELL
THE TUNNEL UNDER THE CHANNEL
ALONE THROUGH THE DARK SEA
AN AGENT IN PLACE
THE WITHERING RAIN
COMPUTER CAPERS

The Pendulum and the Toxic Cloud

THE COURSE OF DIOXIN CONTAMINATION

THOMAS WHITESIDE

YALE UNIVERSITY PRESS, NEW HAVEN AND LONDON, 1979

Nearly all of the text of this book appeared originally in *The New Yorker.*

Designed by Sally Harris
and set in Linotype Times Roman.
Printed in the United States of America by
Vail-Ballou Press, Binghamton, N.Y.

Published in Great Britain, Europe, Africa, and Asia (except Japan) by Yale University Press, Ltd., London. Distributed in Australia and New Zealand by Book & Film Services, Artarmon, N.S.W., Australia; and in Japan by Harper & Row, Publishers, Tokyo Office.

Library of Congress Cataloging in Publication Data

Whiteside, Thomas, 1918–
The pendulum and the toxic cloud.

(Yale fastback ; 20)
Most of the text appeared originally in the New Yorker.
Bibliography: p.
Includes index.
1. Tetrachlorodibenzodioxin—Environmental aspects.
2. Seveso, Italy. I. Title.
QH545.T44W47 1979 615.9'51'1 78-65495
ISBN 0-300-02274-3
ISBN 0-300-02283-2 pbk.

To
William Shawn

Contents

1

YEARS OF HERBICIDAL ADVENTURISM

ON APRIL 15, 1970, Dr. Jesse L. Steinfeld, the Surgeon General of the United States, appeared before a hearing of the Senate Subcommittee on Energy, Natural Resources, and the Environment, headed by Senator Philip A. Hart, which was investigating the potential hazard to human health of the widely used phenoxy herbicide 2,4,5-T. He announced a series of governmental actions aimed at limiting the use of 2,4,5-trichlorophenoxyacetic acid around homes, on farms, and on other areas in the United States. On the same day, the Deputy Secretary of Defense, David Packard, announced that the American armed forces, which had been using 2,4,5-T in the Vietnam war, would stop using it. The employment of the compound in aerial defoliation operations over huge tracts of South Vietnam—for the stated purpose of denying cover to Vietcong forces but also for the unstated purpose of creating a flow of refugees from villages controlled by the Communists into areas controlled by the South Vietnamese government—had aroused increasing protests from biologists in this country, who maintained that the 2,4,5-T had a potential for causing birth defects among the offspring of Vietnamese women exposed to it. Indeed, studies made for our government from 1966 onward by an organization known as Bionetics Research Laboratories, of Bethesda, Maryland, had shown that the herbicide caused a significant number of deformities in the unborn offspring of female mice and rats. Officials of the Dow Chemical Company, the principal

American manufacturer of 2,4,5-T (it is also made at this time by Transvaal, Inc., of Jacksonville, Arkansas, and the Thompson-Hayward Chemical Company, of Kansas City, Kansas), claimed that the Bionetics findings were invalid, on the ground that the samples of the chemical used in the experiments had contained uncharacteristically high levels of a toxic contaminant, 2,3,7,8 tetrachlorodibenzo-*p*-dioxin—commonly referred to as TCDD, or, by chemists familiar with the subject, either as tetra dioxin or simply as dioxin. The government put off taking any decisive action. In 1970, however, just as the Hart hearings got under way, another study made under federal auspices, but conducted by the National Institute of Environmental Health Sciences and using samples of the substance that were far less heavily contaminated with dioxin, still showed that 2,4,5-T had significant teratogenic (fetus-deforming) effects on mice. It was largely on the basis of this second study that the new restrictions announced by the Surgeon General were placed on domestic use of the substance.

These government actions, together with the undertaking by the American military to end the employment of 2,4,5-T in Vietnam, thus putting an effective stop to most herbicidal warfare operations there, seemed to most people to have taken care of the problem, and after a while the matter dropped from public sight.

Yet after all these years, 2,4,5-T is still being used extensively in this country. In fact, the restrictions imposed in 1970 appear to have affected less than twenty per cent of the actual use of the herbicide in the United States. Its largest producer is still Dow Chemical, and the compound is being used—mainly in aerial spraying operations—to defoliate and kill vegetation on power-line, highway, pipeline, and railroad rights-of-way, and to kill shrub and broad-leaved plant life on rangelands, pasturelands, and forestlands. Millions of acres a year are involved, and since the effects of a single spraying continue for years, the total area affected at any given moment is considerably larger than the annual spray rate indicates. In Vietnam, almost five million acres were sprayed between 1962 and 1970. In this country, an area slightly larger than that is being sprayed with the herbicide *each year*. Much of the current spraying is being carried out under the auspices of the federal government itself. Various state governments also contribute to the amount of 2,4,5-T being laid down, through their own spray programs. For some time, the United States Forest Service has been engaged in intensive programs to spray 2,4,5-T on national-forest areas—the announced purpose being to kill broad-leaved vegetation and encourage the growth of commercially valuable softwood trees such as spruce and pine. But not only broad-leaved weeds are the targets of the operations. For example, in the Ozark and St. Francis National Forests, in Arkansas (a state in which 2,4,5-T is being heavily used on rice crops), the Forest Service has been spraying and killing stands of oaks and other hardwoods at a rate of from fifteen thousand to twenty thousand acres a year in order to promote the growth of softwoods. Such spray programs have aroused the strong opposition of many local residents, and spurred action by regional environmental and wildlife organizations. As a result of a lawsuit in an Arkansas federal court by one such group, which charged that the Forest Service had violated the National Environmental Policy Act by engaging in 2,4,5-T-spraying programs without proper authorization, the court in June, 1975, enjoined the Forest Service from continuing the program in Arkansas until it had filed a proper environmental-impact statement. Other lawsuits by environmental groups have brought about similar federal-court injunctions against Forest Service

2,4,5-T programs at one time or another in Michigan, Wisconsin, and Oregon. But there have been no such restraining actions against the use of the product on rangelands and pasturelands. The Dow Chemical people will not say how many tons of their 2,4,5-T are now being used in such applications; they consider this to be privileged corporate information, to which the public is not entitled. Dow is obliged to give its 2,4,5-T production figures to the Environmental Protection Agency on request, yet one cannot get this information from the E.P.A., either. Such figures are classified as trade secrets, which the E.P.A. is forbidden by law to disclose, and so chemical producers are well protected from any outside attempt to learn how much herbicide a particular manufacturer is selling to be used on, for instance, the grazing land of cattle whose meat will wind up on the dinner tables of the public. However, the total area of rangelands and pasturelands treated annually with 2,4,5-T has been estimated by the E.P.A. to amount to two million acres, and the extent of such lands to have been treated since 1970, when the use of 2,4,5-T was supposedly tightly restricted, can be estimated at twelve million acres. On these lands, unknown numbers of successive herds of cattle have been regularly grazed. Considering the propensities that critics say the dioxin contaminant in 2,4,5-T has for persisting, even at diminished concentrations, in the soil in which it was laid down, and the recognized potential for cumulative damage to animal organisms, as shown in laboratory experiments, the use of this herbicidal compound on lands and on vegetation that cattle feed on while they are being fattened prior to slaughter is raising deeply disturbing questions among scientists.

THE scientists' concern is in no way allayed by the residual effects of herbicidal warfare in Vietnam, in which 2,4,5-T was so widely employed—effects of such an order that some of them are even now clearly visible as detailed from space by the Landsat research satellite. The criss-cross spray patterns left by C-123 cargo planes in herbicidal operations during the nineteen-sixties show up as dismal brown areas, like threadbare sections of felt. According to current scientific estimates, areas badly defoliated by 2,4,5-T may not be restored to their former condition for as long as a hundred years. One cannot help recognizing that such agents are essentially biocides. The scientific history of the development and formulation of 2,4,5-T, from the end of the Second World War to the end of the Vietnam war, is one of reckless effort and grossly inadequate testing. For example, the American military, having developed 2,4,5-T as part of its biological warfare program in the years following the Second World War, unhesitatingly employed it during the war in Southeast Asia, spraying twenty thousand tons over both populated and unpopulated areas of South Vietnam, without the Pentagon's scientists' ever having taken the precaution of systematically testing whether the chemical caused harm to the unborn offspring of as much as an experimental mouse. Now, extensive research, carried out largely in the wake of protests by a small group of scientists, has been done on 2,4,5-T, and as more has become known about its dioxin contaminant the questions about its mode of action and the consequences of its use have become increasingly serious. These questions were underlined by a number of disturbing developments, not the least of which were the aftereffects of an explosion in July, 1977, at a factory outside Milan, Italy, which released a toxic cloud containing undetermined amounts of dioxin over an area in which thousands of people were living.

As for the government restrictions on 2,4,5-T, which most people might assume had effectively disposed of the issue years ago, the regulatory process that is supposed to govern the use of the herbicide can be described as almost stalled, having been impeded by disagreement among scientists, by the determination of the chemical-manufacturing industry to continue production and sale of the herbicide, by bureaucratic backing and filling, by a lack of federal funds for appropriate research, by political pressures, by postponed hearings, by a court injunction obtained by one producer against the Environmental Protection Agency, and by the government's own indecisiveness. The Air Force, which until the spring of 1977 had in storage huge stockpiles of a dioxin-contaminated 2,4,5-T mixture that had been withdrawn from Vietnam, waited years to get a firm decision out of the E.P.A. on which of several elaborate methods would be acceptable for the disposal of its poisonous hoard.

THE restrictions announced by the government in 1970 included the immediate suspension by the Department of Agriculture of manufacturers' registrations (in effect, marketing licenses issued by the government) to sell 2,4,5-T in liquid form for use around the home and to sell any formulation of it used for killing vegetation around lakes, ponds, and irrigation ditches. At the same time, the Department of Agriculture also announced that it intended to cancel the registrations of nonliquid formulations of the herbicide for use around the home and on food crops. These decisions were, as might have been expected, not well received by any of the big manufacturers of herbicides. In May, 1970, a month after the restrictions were announced, Dow Chemical was joined by Hercules, Inc., and Amchem Products, Inc., then also vendors of 2,4,5-T, in appealing the decision of the Department of Agriculture to cancel the product's use on domestic rice crops. The appeal was made on the ground that 2,4,5-T did no harm an associate of Ralph Nader (he is now an executive in the Office of Management and Budget in the Carter Administration), formally demanded, on the basis of available laboratory evidence, that the Department of Agriculture place a full-suspension order on the use of 2,4,5-T on all food crops, and that it restrict other uses of the herbicide. The department refused to do so, whereupon Wellford, who was joined by Consumers Union, took the matter to the District of Columbia circuit of the United States Court of Appeals, asking that the government be ordered to protect the public health by putting the further restrictions into effect. While the Wellford suit was pending, the newly formed Environmental Protection Agency took over the regulation of pesticides and herbicides from the Department of Agriculture, which had long been under severe criticism for laxness in the area of pesticide control. The Court of Appeals held that the department, in its orders concerning 2,4,5-T, had not given sufficient importance to possible hazards to human health in the permitted uses of the herbicide—in particular, to possible hazards that farm workers might be subjected to in its use on food crops—and on January 7, 1971, the court ordered the E.P.A. to reconsider the Department of Agriculture's refusal to place firmer restrictions on the use of 2,4,5-T. In March, 1971, the E.P.A. announced that it would not take wider regulatory action on 2,4,5-T until a science advisory panel it had set up to evaluate the safety of the herbicide had reported to it. Two months later, the science panel found, in effect, that 2,4,5-T did not create any significant health hazard and proposed that the ban on the use of the herbicide around homes be lifted, although it acknowledged the chemical's teratogenic potential by recommending that labels on containers warn, "This compound may be dangerous to pregnant women." The panel recommended that the registrations be restored, provided only that certain limitations be placed on the maximum dioxin content of future 2,4,5-T; it also recommended further research on the potential of dioxin to accumulate in food chains. In a dissenting opinion, Dr. Theodor Sterling, who is currently a professor of computer science at Simon Fraser University, in British Columbia, asserted that the report of the panel had been "overoptimistic in assessing the implications of data" available to it on 2,4,5-T; he characterized the teratogenic potential of domestic uses of 2,4,5-T as "uncertain."

The representatives of environmental groups who had objected to the use of 2,4,5-T were naturally dismayed by the advisory panel's report. They held a press conference at which they criticized the panel's findings and recommendations. Several well-known scientists who had serious doubts about the safety of 2,4,5-T and had been speaking up on the subject—they included the biologists Matthew S. Meselson, of Harvard; Barry Commoner, of Washington University, in St. Louis; and Samuel S. Epstein, of Case Western Reserve University—declared that they found the report less than convincing, mostly because the extent to which the dioxin contaminant in 2,4,5-T might accumulate in the environment and in the human food chain had not been fully explored. Such statements caused the E.P.A. people to regard the recommendations of the advisory panel with some caution. The E.P.A. called upon a group of scientists at the Food and Drug Administration to comment on the report. The F.D.A. people also considered the report less than convincing, and recommended that the existing restrictions on the herbicide remain in force. In effect, William D. Ruckelshaus, who was the Administrator of the E.P.A. at the time, decided to sidestep the cumbersome regulatory machinery whereby the findings of the advisory panel might be made binding pressures with the most basic considerations of human safety." Referring pointedly to "interlocutory judicial jousting which experience has taught us can go on for years," the court reversed the Arkansas district-court decision and upheld the E.P.A.'s appeal. However, Dow was not altogether the loser, because, by the court injunction the company had obtained against the E.P.A., the agency had been effectively precluded for nearly two years from moving forward with its hearing on the safety of 2,4,5-T.

IN the meantime, significant new laboratory evidence had come to light concerning the aftereffects of the use of 2,4,5-T in Southeast Asia. The biologist Matthew Meselson and a chemical associate, Dr. Robert Baughman, had together developed a refined analytical system for detecting the presence of dioxin in parts per trillion, rather than in the parts per billion ratio previously detectable. Then Meselson and Baughman found residues of dioxin—which could reasonably be assumed to have their origin in the earlier 2,4,5-T defoliation operations in Vietnam—in fish and crustaceans that were caught for human consumption in Vietnamese waters; this suggested that the dioxin may have entered into the food chain, probably through contaminated particulate matter that had been swept in suspension down Vietnamese streams and rivers. The finding appeared to some E.P.A. people to indicate a potential threat to health and the environment in the United States itself from even non-cropland uses of 2,4,5-T.

In July, 1973, the E.P.A. announced that it intended to go ahead with its previously proposed hearing to determine whether all uses of 2,4,5-T in this country should be cancelled. The hearing was scheduled for April, 1974, and the agency started on a series of pre-hearing conferences with representatives of the chief interested parties, including Dow, the Department of Agriculture, and such environmental groups as Consumers Union and the Environmental Defense Fund. As time went on, E.P.A.'s resolution to push ahead with regulatory action seemed to falter. William Butler, counsel for the Environmental Defense Fund—who, since he had been representing a public-interest law organization, had what was obviously a partisan interest in the proceedings—gave me this account of what he saw: "As the time for the hearing approached, E.P.A. got more and more shaky as to what the evidence against 2,4,5-T was. The agency had only one lawyer on the case, and one, less than full-time, staff scientist preparing the evidence. The outside interested parties such as the environmental groups had formidable opposition. Dow had enormous scientific and financial resources at its disposal. Although we could consult with people like Meselson, Baughman, and Epstein, we had only one part-time scientist of our own to work on a very complicated issue. As we did our best to pursue the subject, Dow tried to bring as broad a coalition as possible against us, including the Department of Agriculture and the Department of Transportation. The Agriculture people were solidly ranged against E.P.A. The Dow people wanted the Department of Transportation to testify about the importance of using 2,4,5-T to clear railroad and highway rights-of-way. Unfortunately for Dow, the Department of Transportation had a couple of people in the office of its general counsel who knew something about 2,4,5-T. These people had a study made of the experience of several state transportation agencies. The agencies who were questioned on the subject minimized the need for using 2,4,5-T to clear rights-of-way."

About a month before the E.P.A. hearing was to be held, "Dow and the Department of Agriculture put together what I can only call a sham conference on 2,4,5-T," Butler continued. "It was supposed to bring together the biggest experts on 2,4,5-T and produce the definitive word on the subject. However, it turned out that the invited parties were mostly Dow employees and U.S.D.A. folk, who proceeded to give as formal papers what in essence was their forthcoming testimony at the E.P.A. hearings. We suspected that they were doing this so that if there were any conflicts in their proposed testimony these conflicts could be ironed out before the people involved went into the hearings. In other words, it was a dry run to make everybody feel comfortable—a psychological ploy by Dow's attorneys, in my opinion. The Environmental Defense Fund, although an interested party, hadn't been invited to participate. But we found out about the meeting and did manage to get the scientist who was working for us in as an observer."

In June, 1974, the E.P.A. announced that it was withdrawing its order of intent to hold the hearing. At the same time, it withdrew its cancellation order on the use of 2,4,5-T on rice crops. In an accompanying statement, the agency declared that it needed more time to collect evidence concerning the precise effects of residues of 2,4,5-T on the environment and on the biological chain. The environmental groups were outraged. "E.P.A. just got cold feet," Butler said. "They were saying, in effect, that the agency would assume the burden of proving, through further research, that 2,4,5-T was harmful, rather than requiring Dow to show that the herbicide was harmless."

Another man who was close to the situation told me, "I believe that political considerations were involved in the decision. E.P.A. had cancelled uses of DDT in 1972, and was under heavy retaliatory pressure from Representative Jamie Whitten, of Mississippi, who was chairman of the

House appropriations subcommittee that oversaw the E.P.A. budget. DDT was the big pesticide being used on cotton crops and soil, and Whitten was violently opposed to its cancellation. He threatened the E.P.A. men unmercifully. Also, about that time E.P.A. was preparing to open suspension hearings on aldrin, which was the big corn pesticide, and on dieldrin, which formerly had been used to control the boll weevil and now was being used on termites and other pests. Cotton and corn are the crops that account for almost half of the insecticide use in this country. The Department of Agriculture was worried about the implications of regulatory action by E.P.A. on a broad range of herbicides and pesticides. The 2,4,5-T case was considered crucial in the area of herbicides. The top people in Agriculture were strongly opposed to severe regulatory action on any herbicide, fearing a domino effect. They took the position 'We've got to stop this now.' And they were stirring up their legislative friends, who were giving E.P.A. a hard time. And, during all this, Russell Train, who had taken over from Ruckelshaus as Administrator of E.P.A., was under tremendous pressure from the Nixon White House to go easy on regulatory matters. It was not a promising situation in which to break new ground in regulating the use of these chemicals."

The E.P.A. people maintain that the agency called off the 2,4,5-T hearing on the basis of prudent judgment exercised under difficult circumstances. "The problem had to do with the nature of the evidence available against the behavior of 2,4,5-T in the environment," a senior scientist at E.P.A. told me. He went on, "Meselson and Baughman had come up with their new analytical procedures for detecting residues of dioxin in the parts-per-trillion range, and we had spent the best part of a year getting samples of various kinds out of areas in which 2,4,5-T had been arising out of the vehement opposition of certain members of the scientific community to the Vietnam war. In addition, the Dow people maintain that speculation about alleged dangers of 2,4,5-T to human fetuses is unsupported by solid evidence, and they insist that dioxin is a "weak teratogen." They readily concede the great toxicity of dioxin, but they point to the low levels at which, as 2,4,5-T is now formulated, it is present in the herbicide. Further, they assert, after the 2,4,5-T is applied to vegetation the dioxin is broken down by sunlight and bacterial soil action. And they say that during the period before it is thus decomposed it does not readily leach out of soil, and, being almost insoluble in water, does not tend to be taken up in watercourses. Lastly, they say that the concentrations at which dioxin exists in commercially manufactured 2,4,5-T are so minute that when they are further lowered by the herbicide's being sprayed (and thereby diffused) over very large land areas, the risk to human health is infinitesimal, and is not a practical consideration.

Dow's position on the safety of 2,4,5-T is amplified in a vast quantity not only of scientific literature but also of press releases and promotional material put out by the company's press-relations people. In contrast to the pinchpenny P.R. routine of the environmentalist groups critical of 2,4,5-T (the Environmental Defense Fund people regretfully had to levy a charge of ten cents a page on Xerox copies of scientific papers they passed on to me), the Dow P.R. material is handsomely bound, with expensive paper, elegant design, and, sometimes, four-color illustrations. The Dow publications include a "Press Reference Manual" entitled "The Phenoxy Herbicides," with individual chapters printed on paper of different colors—white, green, yellow, blue, gold, pink, and tan. A preliminary note explains that this press kit was prepared by the Wisconsin Agri-Business Council and reprinted by Dow. The Wisconsin Agri-Business Council is a trade association of which Dow is a member, and the press kit was prepared on behalf of the council by a firm that had done P.R. work for Dow in the past. Then, there is a luxuriously printed publication called "Chemicals, Human Health and the Environment: A Collection of Dow Scientific Papers." In addition, Dow has put out spiral-bound company rebuttals to some of the printed criticisms of the continued commercial use of 2,4,5-T; these Dow publications quote passages of the criticism along with answering comments from Dow. For the agribusiness community, Dow issues a four-color periodical publication, called *Industrial Vegetation Management,* which contains articles with titles like "Herbicide Use Saves Energy." Its article "South Dakota State Highway Maintenance Management System" is a description of the use of Dow herbicides along the grassy edges of highways. All this is presumably intended to forward a principle enunciated by Dow as company policy and cited in *Industrial Vegetation Management* in connection with a chronology of federal regulatory activity affecting 2,4,5-T: "Countervailing forces must be applied to areas which have been subject to overwhelming emotional pressures of only one kind, so that the environmental pendulum will be swung from environmental extremism to reasoned center and we may all move forward constructively to deal with the major tasks at hand."

In this, Dow has been joined by the Council for Agricultural Science and Technology, an organization that early in 1975 issued a report on the use of phenoxy herbicides characterizing the debate over the use of 2,4,5-T in Southeast Asia and elsewhere as a "largely political controversy." Concerning the controversial subject of the dioxin contaminant in 2,4,5-T and its bioaccumulative properties in algae, snails, and fish, the report declared this to be "of no practical interest," since there was "no substantial supply of the chemical in the environment subject to accumulation," and further stated, "It is quite literally an academic problem in the sense that many questions remain unanswered about TCDD, and research should be continued until all aspects of the toxicity, occurrence, and environmental fate of this chemical are thoroughly understood." In the concluding passages, the council declared that any reduction in the use of the phenoxy herbicides (including 2,4,5-T) on farms, pastures, and rangelands would probably require about twenty million extra man-hours of labor annually to maintain production at the current levels, and it would bring about increases in food costs to consumers. To judge from its tone, the report could have been written at Dow headquarters in Midland, Michigan. And, indeed, the council report was extensively quoted in the Wisconsin Agri-Business Council press manual that was reprinted and circulated by Dow. It seems that the whole American agribusiness establishment—the Department of Agriculture included—has formed a solid front on the question of 2,4,5-T.

Taking into account the contributions that Dow had made to the war effort in South Vietnam with its vast supplies of 2,4,5-T and other chemicals used there, some people might suppose that the Department of Defense would heartily endorse the unequivocal position of Dow and its sympathizers concerning the alleged innocuousness of 2,4,5-T. But this has not been altogether the case. The attitude toward 2,4,5-T has somewhat changed at the Department of Defense, it seems. The days are over when, during the height of herbicidal operations in Vietnam, an American military spokesman could declare that the relevant herbicidal agents were "not at all harmful to humans or animals"—and illustrate this statement by dabbing on his tongue a bit of liquid from a bottle that sat on his desk. The facts put before the Hart subcommittee concerning the fetus-deforming effects of 2,4,5-T in tests on laboratory animals not only brought to a halt, in 1970, the military use of 2,4,5-T in Vietnam but formed part of the background of an official declaration by President Gerald Ford, early in 1975, that thenceforward the United States would make no first use of military herbicides in offensive operations.

IN 1971 and 1972, the military had withdrawn huge stocks of Agent Orange—a fifty-fifty mixture of 2,4,5-T and 2,4-D, which was principally used in its herbicidal operations—from Vietnam, and the Air Force placed in storage on Johnston Island, in the Central Pacific, one million four hundred thousand gallons of Agent Orange in more than twenty-five thousand fifty-five-gallon steel drums. In addition, more than fifteen thousand other drums, containing eight hundred and sixty thousand gallons of Agent Orange, were stored at the Naval Construction Battalion Center at Gulfport, Mississippi. With the passage of time, the Air Force people in charge of these stockpiles became increasingly concerned about finding a safe way of disposing of all that Orange, much of it contaminated at relatively high levels with dioxin. Indeed, the matter of disposal became an urgent problem, because the drums in which the mixture was stored were steadily rusting and rotting away and leaking their contents. For some time, the Air Force gave consideration to a scheme under which its Agent Orange would be reprocessed by a private company to make it less toxic and then sold to the Brazilian government for clearing undeveloped agricultural land. But the State Department would not approve this scheme without the prior licensing approval from the E.P.A., and that was not forthcoming, so the scheme fell through. In the meantime, the Air Force had been drafting an environmental-impact statement outlining a variety of means by which Agent Orange stocks might possibly be disposed of without danger to human health or to the environment. The initial method that the Air Force proposed was incineration of the 2,4,5-T at a commercial incineration site in Deer Park, Texas. The Texas Air Control Board did not care for this idea. It protested, "The area around the proposed site of incineration . . . is a highly industrialized area [that already] has relatively high concentration of air pollutants. The addition of combustion products from the incineration of over two million gallons of Orange herbicide into the atmosphere of this area over a prolonged period could compound an existing problem and might very well prove harmful." In February, 1972, the Mississippi Air and Pollution Control Commission formally requested that the Agent Orange stored at Gulfport "be removed from its present location . . . immediately." The letter noted that the 2,4,5-T stocks posed a threat to nearby recreational and shellfish waters, and referred to the area's dense human population and to a "history of hurricanes and tornadoes in that particular section of the country." The State of Mississippi just did not want the stuff around.

The Air Force people had some more ideas for disposing of the Agent Orange. They tried out what they thought was a good one. They wrote to the manufacturers who had turned out most of the contaminated Agent Orange herbicide in the first place, and asked them to take the Agent Orange back. According to a subsequent statement made to the E.P.A. by an Air Force official, this proposal "created what might be known as a wide wave of disinterest" on the part of the manufacturers. The Air Force people considered yet another way of getting rid of the stocks. "The prudent disposition of Orange herbicide for use on privately owned or governmentally owned lands may have a tremendous impact on increasing the availability of . . . rangelands and forests," an Air Force document stated, but it noted, not very hopefully, that any such use "would require [E.P.A.] registration."

The Air Force continued to explore disposal alternatives: the possibility of burying the Agent Orange stocks in deep underground cavities created by nuclear test explosions, or in other underground repositories. But each of these schemes was found to be unsatisfactory. In 1975, the Air Force launched a pilot project in which quantities of Orange were reprocessed in such a way as to extract enough of the dioxin contaminant to make the Orange salable on the commercial market. The extracted dioxin was concentrated in large charcoal-filled steel cannisters. The reprocessing was carried out by a private chemical company in Mississippi, using naval facilities at Gulfport. But then the commercial processor (and, ultimately, the Air Force) was confronted with the problem of how to dispose of the dangerous dioxin residues concentrated in the steel cannisters. The processing company contracted with the owners of a landfill near West Covina, California, to bury the steel cannisters there. The cannisters were sent to California for burial in the landfill, but when this came to the attention of the California State Board of Health and the state's Hazardous Waste Technical Advisory Committee, these agencies refused to have the dioxin remain within the state. The processor then arranged with a corporation called Chem-Nuclear Systems, Inc., to have the deadly material shipped by truck to a hazardous-waste-disposal site in Arlington, Oregon, where it was to be buried thirty feet underground. But when Senator Mark O. Hatfield, of Oregon, learned of this action, he raised an alarm with the U.S. Environmental Protection Agency and with the Air Force, and asked that the dioxin be placed under military control and shipped out of Oregon. Eventually, this was done, and the dioxin cannisters were flown to McChord Air Force Base, in Washington, for special transshipment by military cargo plane to Johnston Island. By mid-1976, the Air Force people gave up the Orange-reprocessing scheme as one that, in effect, represented a cure potentially even more dangerous than the disease it was aimed at. Eventually, the Air Force came to the conclusion that the best way of disposing of the Agent Orange stocks would be one that they had been considering for some time, but had been unable to get authorization from civilian agencies to carry out—that is, through incineration at sea, on a German-built ship, owned by a company in Singapore, that had been designed for just such a purpose. The ship, named the Vulcanus, would carry the 2,4,5-T stocks from Gulfport and Johnston Island to a remote area west of Johnston Island, and the incineration would be done there. The departure of the Vulcanus, loaded with Agent Orange, from Gulfport and then from Johnston Island for the remote incineration site was to be carried out, according to the Air Force, with "maximum possible" precautions. The fifty-five-gallon drums, once emptied of Agent Orange, were to be thoroughly rinsed, then mechanically crushed, and the crushed metal was to be kept in an isolated place until it was transported to steel mills to be smelted down at a temperature of approximately two thousand nine hundred degrees Fahrenheit, which was calculated to assure proper destruction of any residues of dioxin.

The Air Force people, when they had originally drawn up plans late in 1974 for their Pacific incineration scheme, had hoped to put it into effect fairly rapidly, but the E.P.A. had objected, on the ground that the Air Force needed an ocean-dumping permit, which the agency hesitated to grant.

Meanwhile, the storage drums of Agent Orange—particularly those on Johnston Island—were deteriorating so steadily that an average of five drums a day were springing leaks. Some five thousand drums, containing more than a quarter of a million gallons of Agent Orange, had rotted through in this way. In their eagerness to get rid of their Agent Orange stocks, and frustrated at the delays in getting E.P.A. permission to proceed with their incineration scheme, some of the Air Force people at the Pentagon began speaking about 2,4,5-T and environmental safety in the earnest tones that one might expect to hear from officials of the Environmental Protection Agency, while at the same time some of the E.P.A. people went on regarding the problem in the non-urgent manner adopted in former times by people in the Pentagon. Only in April, 1977, seven years after the use of Agent Orange by the military in Vietnam was stopped, did the E.P.A. issue a permit for the incineration at sea of the remaining stockpiles of the contaminated herbicide. In June, the Vulcanus, loaded with eight hundred and sixty thousand gallons of Orange, sailed from Gulfport toward the Pacific Ocean incineration site. The incineration of the huge quantity of Orange stored on Johnston Island was to follow.

THE argument of the Dow people, reinforced by their research into the subject, that 2,4,5-T is being sprayed or otherwise laid down for various domestic purposes in such a widely dispersed fashion as to deposit almost meaninglessly minute traces of dioxin has been all but unanimously accepted in the agricultural-chemical industry, and has impressed to a considerable degree some of the principal scientists within the E.P.A. Applied according to instructions, 2,4,5-T should in fact deposit only low-level residues of dioxin which then are subject to biodegrading action by sun and by soil bacteria. The dioxin that may linger in the environment, therefore, exists at such low levels, in the view of the manufacturers, as to remain far below the threshold of conceivable menace to the health of human beings or animals. But those skeptical of the supposed innocuousness of these attenuated herbicidal traces insist that dioxin is no ordinary poison, and that not nearly enough is known about its exact potency and persistence, except that it is without question one of the most toxic substances known to man; indeed, it is exceeded in toxicity only by the botulinum toxin and by certain substances in the armamentarium of nerve gas and other chemical-warfare weapons. The capacity of low dosages of the dioxin for killing or damaging, in plainly observable ways, the fetuses of experimental animals is now well established. What is not yet properly understood is the potential of dioxin to exert long-range effects, not readily observable, on internal organs, and on the central nervous and immunosuppressive systems of animal organisms, and even on the genetic chain.

The power of dioxin to exert harmful effects on human beings as well as upon experimental animals is indicated by an occupational illness, chloracne, that has attacked workers in 2,4,5-T factories—and even people in contact with such workers—in the past. This illness, as it had occurred among a number of Dow employees working with wastes from 2,4,5-trichlorophenol—a chemical that is a precursor of 2,4,5-T—which contained anywhere from twelve hundred to two thousand parts per million of dioxin, was described before the Hart subcommittee in 1970 by Dr. Julius E. Johnson, a vice-president of Dow Chemical and the company's director of research and development, as "a skin disorder mostly prevalent on the face, neck, and back. It is similar in appearance to severe acne often suffered by teen-agers." As Dr. Johnson explained the situation to the senators, chloracne might seem to his listeners a transient cosmetic problem, perhaps more annoying than serious. But that was not the case at all, as the Dow people were only too well aware.

Long before the advent of 2,4,5-T—in fact, since the mid-nineteen-thirties—the Dow people had known that various polychlorinated derivatives of chlorophenols (including fungicidal preparations, such as pentachlorophenol, that were used for treating wood, paper, paints, and other products and that Dow sold under the trade name of Dowicides) had produced chloracne-like symptoms among workers exposed to them. Dow's 2,4,5-trichlorophenol appears to have been no exception. In 1964, when Dow's production of 2,4,5-T was rapidly expanded to provide Agent Orange for Vietnam, more than seventy workers at a Dow 2,4,5-trichlorophenol factory developed cases of chloracne, a dozen of them severe. Two months before Dr. Johnson described the symptoms of chloracne to the Hart subcommittee as though the disease were something less than a serious problem, Dr. Benjamin Holder, the director of the medical department of Dow's Midland Division, had informed a group of government chemists that the early symptoms of chloracne included fatigue, lassitude, and depression, and that early signs included the appearance of comedones (blackheads) on the face and body and weight loss, and he had reported that heavy exposure to dioxin-contaminated trichlorophenol waste caused damage to internal organs and disorders of the nervous system. Workers at other companies producing 2,4,5-T had also been subject at various times to outbreaks of chloracne. Back in 1949, at a 2,4,5-T plant owned by the Monsanto Company in Nitro, West Virginia, two hundred and twenty-eight workers developed chloracne as a result of an explosion there. Their symptoms included skin eruptions, shortness of breath, intolerance of cold, palpable and tender liver, a loss of sensation in the extremities, damage to peripheral nerves, fatigue, nervousness, irritability, insomnia, loss of libido, and vertigo.

In 1953, in West Germany, an explosion occurred in a 2,4,5-T factory at Ludwigshafen am Rhein owned by Badischer Anilin & Soda-Fabrik (B.A.S.F.). The explosion was the culmination of conditions associated with months of overheating occurring in a plant reactor containing 2,4,5-trichlorophenol. During these months, all the male workers and many of their wives and children, and even their pets, had developed chloracne. After the explosion, some workers were found to have suffered serious damage to internal organs, including severe liver damage. Other symptoms included raised blood pressure and myocardial degeneration, severe depression, and disturbances of memory and concentration. One death from intestinal sarcoma was attributed to the poisoning, and fifteen years after the accident the chloracne and its symptoms were reported to be still present in some of the workers, in spite of continued treatment and no further contact with the source of the illness, which was certainly dioxin contamination in the 2,4,5-trichlorophenol being produced in the reactor.

In 1963, in Amsterdam, a similar explosion occurred in a 2,4,5-T factory owned by Philips Duphar, a subsidiary of N. V. Philips of the Netherlands. Fifty workers subsequently came down with chloracne, and many of these suffered damage to internal organs and serious psychological disturbances. The 2,4,5-T plant was closed for ten years, but even at the end of that period the premises were found to be so seriously contaminated with dioxin that the entire plant had to be dismantled, brick by brick, and the contaminated material embedded in concrete, loaded, at a specially constructed dock, on ships, and dumped at sea, in deep waters near the Azores.

In Czechoslovakia, between 1965 and 1969, extensive leaks occurred in the trichlorophenol-processing area of a 2,4,5-T-manufacturing plant near Prague, and a number of workers developed the symptoms of chloracne, including, according to an account in a Czechoslovakian medical journal in 1973, "somatic, mental, and sexual" complaints and considerable weight loss. For most of those affected, the disease reached its peak somewhere between one and two years after initial exposure, but in some patients the symptoms were still manifest eight years after their initial exposure.

In 1968, an explosion took place at a factory in Bolsover, Derbyshire, England, owned by Coalite and Chemical Products, Ltd., which was producing trichlorophenol for both 2,4,5-T and the bacteriostatic compound hexachlorophene. Seventy-nine workers developed chloracne as a consequence of this exposure to tetra dioxin. Not only workers but also members of their families having contact with them became sick. Several of the workers died, the immediate cause of death being severe liver damage. Because of the persistence of heavy dioxin contamination within the factory premises, all factory equipment considered seriously affected (except for some large tanks, which were thoroughly cleaned) was removed and buried a hundred and fifty feet down in a disused coal mine.

EVIDENCE of serious dioxin contamination outside the premises of factories in which 2,4,5-T or the trichlorophenol intermediate material was being produced was also showing up in various places. In 1971, incidents occurred at three horse arenas in eastern Missouri, near St. Louis, in which forty-three horses, scores of chickens, and uncounted hundreds of birds, rodents, cats, and dogs died. An investigation by the Center for Disease Control of the Public Health Service disclosed that a short time before the incidents the riding arenas at a horse-breeding farm had been sprayed with waste oil to keep down dust. Of eighty-five horses that had been exercised in the sprayed arenas, fifty-eight became ill and forty-three died. Among pregnant mares that had used the arenas, there were twenty-six known abortions. A report of the Center for Disease Control stated that many foals that had been exposed only *in utero* died either at birth or shortly thereafter; six of the foals had manifest birth deformities. Further, four children who had played in the sprayed arenas were taken ill. Two of the four had symptoms, from which they fortunately recovered, identical with those of industrial chloracne.

The investigation revealed that the oil had come from a salvage-oil company. Analysis of residues of the sludge sprayed on the arenas showed a high concentration of 2,4,5-trichlorophenol, and pointed to dioxin contaminant in the chemical as the villain. In the sludge that had been collected by the salvage-oil company, it appeared, was waste from a reactor at a factory near Verona, Missouri, which had formerly produced 2,4,5-T for use in Agent Orange. After the use of Agent Orange in Vietnam was brought to a halt in 1970, the factory had been sold. The new owners had gone on making trichlorophenol, using it, as was widely done, as a starting material in the production of hexachlorophene. The reactor's highly toxic dioxin-contaminated waste, which the former owners had had transported by tank truck several hundred miles for disposal by incineration, had been disposed of simply by paying the salvage-oil company to haul it away; the salvage-oil company trucked quantities of it to St. Louis and dumped it into a large holding tank, along with other waste oil and sludge. Some of the mixed waste had been drawn off from the tank for spraying the horse arenas.

2

DISASTER AT SEVESO
JULY, 1977

INITIAL EFFECTS OF THE *ICMESA* EXPLOSION OF 1976

AGAINST the background of all the industrial accidents and other grim episodes, one might suppose that the unusually grave responsibilities attached to the handling of dioxin would have been only too apparent to any company that took the risk of inflicting it, in even the most minute quantities, on susceptible humanity. But in 1976 this proved, on the most appalling scale since Vietnam-war days, not to be so. On July 10, 1976, another explosion—one whose consequences were to affect the lives of thousands of people in frightening fashion—occurred at a chemical factory in Lombardy, Italy, about thirteen miles north of Milan, which was producing 2,4,5-trichlorophenol. The factory is owned by a company called ICMESA (an acronym for Industrie Chimiche Meda Societa Anonima). ICMESA is owned, in turn, by a Swiss corporation called Givaudan, which is a large supplier of cosmetic products and hexachlorophene. The Givaudan corporation is owned, in *its* turn, by the huge international drug-manufacturing company Hoffmann-La Roche, which also has its headquarters in Switzerland.

The ICMESA factory explosion took place in a reactor that was being used to produce 2,4,5-trichlorophenol by combining tetrachlorobenzene with sodium hydroxide in an ethylene-glycol solvent. Normally, the factory completed one such reaction, and also completed a subsequent distillation process to remove the ethylene-glycol solvent, each day. However, for some reason, at the end of the last working shift of the week, in the early hours of Saturday, July 10th, the distillation stage, which routinely completed the production process, was not undertaken and was deferred by the management until the beginning of the next work week. Because of the weekend, only a few watchmen and maintenance people were in the plant when, on Saturday at about 12:40 P.M., the chemical mixture in the reactor, which had become seriously overheated as a result of continued activity, exploded. The force of the explosion blew out a safety disc at the top of the reactor and drove the boiling mixture up a venting pipe leading directly to the open air. The explosion was heard by residents in the area as—in the words of one woman—"a deep rumbling sound." From the vent on top of the building a white plume, in the shape of an inverted cone that was tilted to one side, shot into the atmosphere to a height of about a hundred and fifty feet. In a few minutes, the plume extended to form a white cloud, which drifted in a southerly direction and, upon cooling, descended and settled over a cone-shaped area that included at its apex part of the southern section of the community of Meda and a substantial section of the adjacent community of Seveso. In part dispersed by light northwesterly airs, it then lifted slightly and rolled on, over the communities of Cesano Maderno and Desio, to the south.

Although much of the seven hundred acres on which the visible part of the cloud seemed to have settled consisted of market-gardening plots and fields on which crops such as grain were raised and on which cows were grazed, the area was inhabited by about five thousand people, most of them artisans from southern Italy (particularly from Sicily) and their families, generally large. Since the explosion took place during the noontime meal period, most of those living in the area were probably indoors when the cloud settled, but a number, including children who were outside playing, were enveloped by the settling cloud, which had a pungent smell. It made people cough, and their eyes smarted. Not long after being exposed to the cloud, many people, and especially those living closest to the ICMESA factory, felt nauseated, and parts of their exposed skin became irritated. Many of those who had been outside quickly developed burnlike sores on their faces, arms, and legs. Adults and children alike developed other symptoms, among them headaches, dizziness, and diarrhea. Within two days, small animals in the area, including rabbits, mice, chickens, and cats, began to sicken, and many died. Birds that had flown through the cloud or had been enveloped in it while they were perched in trees died quickly—many before they could even fly out of the area. Interestingly, except for the more severe skin lesions these symptoms and immediate aftereffects were very much like those that had been reported on different occasions, during the war in Southeast Asia, by villagers and peasants in various populated parts of Vietnam—and even by remotely situated Montagnard tribesmen in the Vietnamese highlands—after defoliation raids in which clouds of Agent Orange had been loosed on them, their homes, and their gardens.

Early on Saturday afternoon, the ICMESA management made its first response to the accident, when one of its chemists arrived at the plant and inspected the reactor. He then went out to a cluster of homes lying on the edge of several fields in the path of the cloud about a thousand feet away from the reactor and suggested to people living there that they refrain from eating vegetables from their gardens. The following day, Sunday, Francesco Rocca, the Mayor of Seveso, received a telephone call at his home (which itself was not in the path of the cloud) from a doctor friend in the area, who told him that the plant manager at ICMESA and one of his assistants were looking for the local health officer but were unable to find him, and that they wanted to talk to Mayor Rocca. The Mayor invited them to his home. Late in October, when I talked with Mayor Rocca at his office in Seveso, he said that the two people from ICMESA had told him that a cloud "possibly containing some material of a herbicidal nature" had escaped from the factory the day before, and that they felt it wise to have people living in the vicinity of the ICMESA factory advised not to eat fruit from their trees. "They said nothing at all about dioxin or TCDD," Mayor Rocca told me. Since the ICMESA plant itself is within the town of Meda, Rocca informed the mayor of Meda, Fabrizio Melgrati, of what he had been told. Signor Melgrati told the local police what was going on. On Monday, Rocca and Melgrati jointly requested the local health officer to inspect the ICMESA reactor from which the chemical cloud had emanated, and, as before, according to Rocca, the ICMESA management said nothing about the danger that dioxin was present in the fallout. On Tuesday, local health officers reported to the two mayors that the leaves of vegetables and of some trees in the area were wilting. Their reports mentioned no damage to human beings. However, that same day local residents brought in several children and showed Mayor Rocca skin lesions that the children had developed. The Mayor called the ICMESA management to arrange a meeting, and on Wednesday two representatives of ICMESA came with a map of the locality, and, indicating on it the area immediately southeast of the factory, they suggested once again that people be advised not to eat vegetables grown there. And once again, according to Rocca, the ICMESA management representatives made no mention of dioxin and said nothing else to indicate that the cloud might contain an extraordinarily dangerous substance. They said that they had sent samples of the material deposited by the cloud to the Givaudan technical headquarters in Dubendorf, just outside Zurich, for analysis and were awaiting the results.

On Thursday, July 15th, five days after the contaminating explosion, Mayor Rocca issued a proclamation warning the population not to eat fruit or vegetables from what was apparently the most strongly affected area. By that time, birds and small animals were dying at a faster rate. And the effects on human beings were becoming more serious. By Friday, nineteen children were hospitalized near Seveso. The condition of four of these children proved to be serious enough for them to be transferred to the Niguarda Hospital, near Milan, for further examination and treatment. Mayor Rocca had another meeting with ICMESA representatives. He told me that at the meeting he had been prepared to order evacuation of the area that seemed most affected but that the ICMESA people told him this would not be necessary, since there appeared to be no hazard to the local people in continuing to live in their own homes. By the end of the day, the Mayor issued another proclamation, warning people to destroy any produce they had picked since the explosion, and he himself went out to warn one family after another in the cone-shaped area where the cloud had first descended, near the ICMESA factory, not to pick, eat, or store their fruit or vegetables.

The Lombardy provincial health authorities seem to have been remarkably slow in getting into action. It was not until Saturday, July 17th, a week after the explosion, that the director of the Laboratory of Hygiene and Prophylaxis of the Lombardy provincial administration, in Milan, visited the ICMESA plant. There the laboratory director, Dr. Aldo Cavallaro, heard company technicians explain that the immediate cause of the explosion was an exothermic reaction, or increase in temperature arising out of chemical activity within the reaction vessel. According to

Dr. Cavallaro's later account, when he attempted to go near the reactor itself ICMESA technicians insistently dissuaded him from doing so, saying that a toxic by-product might be present. However, he has noted, nothing was said about dioxin or TCDD. Upon returning to his laboratory in Milan that night, Dr. Cavallaro consulted all references to trichlorophenol which were available in the technical library at the laboratory, and he learned that elevated temperatures in trichlorophenol manufacture made possible the formation of the extraordinarily toxic TCDD. On Monday morning, Dr. Cavallaro asked the ICMESA management people whether tetra dioxin could, in fact, have contaminated the cloud. When the ICMESA people conceded that it could have, and that the chemists at Givaudan's research headquarters outside Zurich were studying precisely this question, Dr. Cavallaro flew to Zurich the same day to demand further information from the Givaudan management. At Dubendorf, the Givaudan people conceded that the samples sent to them from the ICMESA plant showed some dioxin contamination.

The admission by Givaudan that the chemical fallout was contaminated with one of the most toxic substances known to man thus came a full nine days after the explosion—a circumstance that particularly embittered the feelings of the thousands of people whose lives were ultimately affected by the escaped cloud. Two days after the Givaudan management conceded the presence of dioxin in the area, the managing director and the plant manager of ICMESA were charged under Italian law with negligently causing a disaster, and were placed under house arrest.

It was not until July 24th, two weeks after the explosion, that a physician representing Hoffman-La Roche notified Vittorio Rivolta, the chief of the Lombardy regional health administration, that the dioxin contamination from the cloud was serious enough to warrant evacuation of the population from the area apparently most affected. Plans for undertaking the evacuation and sealing off that area were announced by the provincial authorities the following day. The area involved was divided into two zones. The first, designated Zone A, consisting of two hundred and sixty-seven acres immediately to the south of the ICMESA plant and inhabited by seven hundred and thirty-nine people, was to be entirely evacuated. Zone B, consisting of about six hundred and sixty-five acres, in which a total of about five thousand people were living, was to be sealed off from non-residents, but the residents themselves could continue to live in their homes, although their children under fifteen, who numbered about eighteen hundred, were to be taken out of the zone during the day, so as to cut down their exposure to contamination.

The evacuation from Zone A began on July 26th, past barbed-wire barriers that had been strung around the zone by Italian Army troops, who now guarded and patrolled its perimeter. Between July 26th and August 2nd, about a hundred residents left the zone. On August 2nd, a second large contingent, of four hundred and forty-four remaining residents, also left. In leaving, the inhabitants were officially forbidden to take household goods or any but their most essential personal possessions, and were instructed to leave all food and domestic animals behind. (The animals were to be left in the care of veterinarians.) The evacuees were given allowances by the provincial authorities and were housed in a large, luxurious apartment and hotel complex in Bruzzano, a suburb just north of Milan.

How long the evacuated people were to be kept away from their homes nobody could tell them. By the time of the evacuation, thirty-six people had been hospitalized with skin lesions and other symptoms. Bird life appeared to have been devastated; fields, gardens, and orchards were littered with the carcasses of swallows, martins, warblers, and goldfinches, and also with those of thousands of rats, mice, and moles. Both brown field rabbits and white rabbits that residents of the area had been raising for food had been dying by the hundreds, and chickens by the thousands. Cats that survived were meowing piteously; dogs, which are known to be comparatively resistant to dioxin poisoning, looked sickly, and their behavior was reported to be nervous and aggressive.

To measure and analyze the extent and the effects of the contamination, the provincial authorities set up various committees, composed of toxicologists, chemists, and epidemiological specialists from Milan—primarily from the Mario Negri Pharmacological Institute and the University of Milan's School of Pharmacy, its Institute of Hygiene and its Labor Clinic. When analysis of soil samples taken from the area reached a further stage, the authorities decided to enlarge the danger zone, and Zone A was extended to take in a good part of Seveso; Zone B also was enlarged. A third, outer zone, consisting of several thousand acres and containing some twenty thousand residents, was designated as a precautionary area, in which intensive health monitoring would be maintained.

Through most of the affected region, the only free traffic that continued was that on the autostrada between Milan and the Lake Como region, to the north. The autostrada ran through the eastern part of Zone A and the western part of B. It was bordered on each side with barbed wire, patrolled by armed soldiers, and posted with signs warning motorists to drive with closed windows and at a very moderate speed, so as to cause as little dust as possible. Inside Zone A, the scene was desolate indeed, inhabited only by occasional hooded figures encased in impermeable white decontamination suits and boots and wearing face masks: scientists monitoring soil samples, and veterinarians collecting dead animals in plastic bags. From time to time, shots could be heard as dying animals were put out of their misery and those still capable of moving were killed to prevent them from travelling out of the contaminated zone. Toxicological analysis demonstrated beyond doubt that most of the animals already found dead had died from dioxin poisoning, and post-mortem examination showed extensive liver damage. The livers of dead rabbits showed concentrations of one and a half micrograms of dioxin per gram of liver—a figure indicating that the animals had been exposed to dioxin at concentrations many times the established lethal dose for rabbits in the laboratory.

Forced from their homes, which in many instances they had built themselves or with the help of neighbors, and quite uncertain of what the future might hold for them, the people from Zone A were in a shocked and bewildered state, as were people still living in Zone B. Most of the furniture in Italy is produced at factories in and around Seveso, Meda, Desio, and Cesano Maderno, and the region contains perhaps three thousand artisans' shops, in which, before the ICMESA accident, carpenters and other workers turned out articles for the furniture factories on a cottage-industry, piecework basis. In Zone A, these artisans' shops were completely closed, and so were the furniture factories. Factories in adjacent areas that might have been affected remained open, but when the Italian press and television reported on the seriousness of the effects of the ICMESA explosion and broke the news that the area was contaminated with deadly dioxin, demand for furniture and other goods from Seveso and the region around it dried up.

People did not want to touch furniture that had been made in Seveso. The livelihood of the artisans seemed to be vanishing. For all the attention that was suddenly focussed on the evacuees, they came to feel isolated—regarded by many as pariahs. This feeling of isolation was heightened by the fact that none of the social workers who had now been sent to assist them, and none of the medical specialists and toxicologists who were now examining them, was in a position to tell them what was likely to happen to them, or what their state of health would be, or when, if ever, they might be able to return to their homes and their normal lives in Seveso.

Of all the uncertainties facing the people of Seveso, none caused more agony than that raised by reports in the press and on television, and acknowledged by the regional health authorities, of the fetus-deforming effects of dioxin on laboratory animals. When the toxic cloud descended, several hundred women in the areas now designated Zone A and Zone B were pregnant, and, in addition to their other troubles, they were confronted with the horrifying possibility that the babies they were carrying might be born with some kind of malformation. They were racked by conflicts involving their maternal instinct, traditional religious precepts, and their recognition of the dreadful responsibility of bringing into the world children who might possibly be doomed to a maimed existence. Many of them felt obliged to consider, most reluctantly, whether, instead of inflicting such possible penalties on their offspring, they should resort to abortion. In Catholic Italy, the law had traditionally made abortion a serious crime, but only a few months before the ICMESA explosion the Italian supreme court had handed down a ruling that made it possible for pregnant women who were faced with unusual danger to mother or child to apply for permission to obtain therapeutic abortions, with permission granted only after rigorous inquiry and under extremely strict conditions. In consideration of the special plight of the pregnant women of the Seveso area, the Italian government announced that it would permit these women to have abortions if their applications to do so were approved by their attending physicians on a case-by-case basis.

This move was sternly opposed by the Catholic Church on doctrinal grounds. The Vatican itself plainly expressed its disapproval, and the Archbishop of Milan declared from the pulpit that if children of the pregnant Seveso women should indeed be born with malformations, there would always be good Catholics who would care for them if their own parents did not wish to. But the government's move on abortion was received with approval by the Italian left. Before long, the unfortunate women of the Seveso area were the center of an issue so highly politicized as to make their distress all the greater. Altogether, thirty-four women from the immediately affected area received permission for therapeutic abortions, and twenty-six of them had the operation performed at the Mangiagalli Clinic at the University of Milan and two at a clinic in Desio. But an additional number of the women—as many as a hundred and twenty of them—are believed to have forgone the formalities and the waiting period required, and to have had abortions performed either illegally in the locality or legally outside Italy.

IN accounts of the ICMESA explosion appearing in the Italian press and on television, there was, not unexpectedly, a great deal of severe criticism of the ICMESA management for permitting conditions that made possible the contamination of the area with such a toxic substance as dioxin—notably, the use of a reactor so improvidently designed that in the event of an accident, however unlikely, it would vent dioxin right into the air instead of passing the poison into a sealed container. The ICMESA company and its Swiss owners were also indignantly criticized for their failure promptly to warn the people in the area and the authorities of the true nature and the extreme danger of the toxic cloud. Various other charges were made in the media —for example, that the ICMESA plant had been secretly supplying trichlorophenol for use in 2,4,5-T being acquired by the North Atlantic Treaty Organization. The Givaudan company management has repeatedly denied that either the parent company or ICMESA had any connection with 2,4,5-T production, and, because of the nature of Givaudan's business, there is little reason to doubt this assertion.

Givaudan produces most of the world's hexachlorophene, and the trichlorophenol produced by the ICMESA plant was destined for use as a source material for hexachlorophene. Givaudan also produces a wide variety of aromatic essences that it supplies to the perfume industry, and it was to carry out this business that Givaudan was originally founded, in 1898; the company developed hexachlorophene as a bacteriostatic agent, and got into its manufacture on a commercial scale in the early nineteen-forties. As for ICMESA, it had also been in the aromatic-compounds business for many years, as an independent supplier to Givaudan. Givaudan acquired control of ICMESA in the nineteen-sixties, and took it over formally in 1969, six years after Givaudan itself was bought up by Hoffmann-La Roche. When Givaudan acquired ICMESA, it converted part of ICMESA's production from aromatic compounds to trichlorophenol for shipment to Givaudan's hexachlorophene-manufacturing facilities—in Switzerland and in Clifton, New Jersey, where Givaudan has long maintained a plant. Why should a Swiss corporation producing hexachlorophene, if that corporation decided to engage in the tricky business of making the trichlorophenol within the company's own facilities, buy a factory in northern Italy and make the material there rather than near its own Swiss plant? The answer given in one Hoffmann-La Roche company publication is that "in Switzerland the labor market was so tight, and the number of guest [i.e., foreign] workers so great that it seemed better to create jobs for guest workers in their own countries than to import them to Switzerland." But this explanation, when it was reported in the Italian press, did little to moderate the indignation of critics of the company's behavior before and after the explosion. It is a common contention of these critics that northern Italy has become, in effect, a dumping ground for almost every conceivable kind of industrial pollution, from operations set up by foreign-controlled multinational corporations whose headquarters' countries would be unlikely to tolerate the loose manufacturing standards allowed in Italy. And, indeed, it appears extremely doubtful whether Swiss government regulations would have permitted the manufacture of trichlorophenol under the conditions that prevailed at the ICMESA plant.

The Givaudan management persistently denies that the safety standards employed at ICMESA were lax, or that Givaudan was negligent concerning the problem of dioxin contamination. Givaudan's explanation of how ICMESA came to be involved in hexachlorophene production is of interest, especially since the background of its problems associated with the production of hexachlorophene happens to have a certain relationship to recent developments in the 2,4,5-T market in the United States. Dr. Guy Waldvogel, the general director of Givaudan, told me in an interview in 1977 that by 1974 Givaudan had become concerned about the quantity of dioxin and other impurities in the trichlorophenol that had been supplied to it by independent companies in the United States and elsewhere. "We had two reasons for going into trichlorophenol production on our own—which we did on a regular basis in 1975—instead of continuing to rely on our usual suppliers of it," Dr. Waldvogel said. "One was a worldwide shortage of supplies of trichlorophenol, resulting from the demand for it in herbicide manufacture. The other was the growing difficulty we encountered in obtaining grades of trichlorophenol that met the analytic specifications we had set. We were imposing on ourselves increasingly stringent quality requirements, which were not satisfied by the trichlorophenol material on the market, and we felt we could produce a better grade ourselves, using high quality controls."

Although Dr. Waldvogel did not say so, other important considerations undoubtedly entered into Givaudan's decision to produce its own trichlorophenol. After 1970, when the Hart Senate subcommittee hearings precipitated restrictions on the use of 2,4,5-T and focussed attention on the frightening toxicity of dioxin, the producers of hexachlorophene had become more and more sensitive about the association of their product with trichlorophenol and its dioxin contaminant. Originally, hexachlorophene was conceived of as an externally applied bacteriostatic preparation to inhibit the spread of bacterial infection in skin burns and the like; it had also been used by hospital physicians and other personnel in scrubbing-up routines. Thus, its uses had been fairly limited. But, beginning in the late nineteen-fifties, heavy sales-promotion techniques by companies to which Givaudan acted as a supplier helped to turn hexachlorophene into a common consumer product. The compound also began to be used in hospitals on a very large scale, for all sorts of purposes—often, it appeared, as a sort of chemical mask for poor hygienic discipline by hospital personnel. It became so routine for newborn babies to be washed with a hexachlorophene solution that the stuff seemed to serve almost as a sort of sheep-dip in hospital nurseries. At the same time, toxicological literature was showing, increasingly, that serious penalties could be connected with the indiscriminate use of hexachlorophene: convulsions in some babies; and, in experimental animals given large dosages, paralysis of limbs and damage to brain tissue. Nevertheless, high-pressure television-advertising campaigns went on proclaiming the virtues of hexachlorophene. It was touted as a new miracle ingredient in all kinds of widely merchandised products for consumers: soaps, deodorants, even vaginal aerosol sprays—the last of these uses being one that, to anybody who knows of the association of hexachlorophene and the trichlorophenol on which it is based, and of the trichlorophenol and its teratogenic contaminant, is plainly horrifying to think of. But after a number of health incidents involving hexachlorophene, which culminated in the deaths in 1972 of some three dozen French infants after they were treated in their cribs with a talc preparation that had been accidentally mixed with large amounts of hexachlorophene, the United States Food and Drug Administration forbade the sale of hexachlorophene for any but restricted medical uses. Since then, hexachlorophene has continued to be sold for indiscriminate uses in a number of other countries, including Italy, but in this country its use is now confined pretty much to physicians' and surgeons' scrubbing-up routines, in which the applied solution can be washed off quickly and with proper care.

These are some of the considerations that may well have impelled the Givaudan people to have their trichlorophenol produced to their own standards by their own corporate subsidiary. It is the contention of the Givaudan management that at the ICMESA plant it deliberately chose a production process that would enable the reactor to operate at relatively low temperatures and at normal atmospheric pressure, in order to insure that minimum amounts of dioxin would be formed and to reduce the possibility of explosion. They insist that they are entirely unable to explain how the reaction got out of hand on July 10th. According to Dr. Waldvogel, ICMESA produced a total of a hundred and five tons of trichlorophenol in 1975 and a hundred and thirty tons in 1976 prior to the July explosion. Of this amount, approximately half was shipped to Givaudan's hexachlorophene-processing factories in Switzerland and half to the Givaudan plant in Clifton. Dr. Waldvogel expressed dismay that ICMESA's international shipments of trichlorophenol had become such an issue. "One hundred and thirty tons—that's a laughable quantity of trichlorophenol in terms of the whole market," he told me. "Where we talk of a hundred and thirty tons a year, the herbicide-manufacturing people think in terms of *thousands* of tons a year." He maintained that after refinement of the trichlorophenol at ICMESA to keep the amount of dioxin contaminant to a minimum, the dioxin in the finished hexachlorophene as sold by Givaudan did not exceed twenty parts per billion.

I was surprised to hear him cite this figure, because in 1970, in the course of working on an article about the problems posed by the use of 2,4,5-T, and its dioxin contaminant, I had looked into other uses made of dioxin-associated trichlorophenol and had asked the Givaudan management people in Clifton whether Givaudan's hexachlorophene contained any dioxin. I was told categorically that extensive research by the company had established beyond doubt that it did not. Now, seven years later, I was told that it did, after all. The explanation for this contradictory information probably lies in the fact that in 1970 the analytical techniques for detecting dioxin in parts per billion were not as reliable as they are now. But a contamination of twenty parts per billion in a medical preparation, and with what is one of the most toxic substances known to man, surely cannot be reckoned as negligible. (In May, 1975, the *Archives of Neurology,* published in Chicago, contained a report of a study based on autopsy reports of premature infants who had died of various causes shortly after birth, and all of whom, at birth, had been routinely bathed in hexachlorophene solutions. The article, by Drs. Robert M. Shuman, Richard W. Leech, and Ellsworth C. Alvord, Jr., reported a high incidence of mild to severe lesions in the brain stems of the deceased infants, the severity of this damage being associated with the number and concentration levels of the hexachlorophene baths the infants had been subjected to after birth.) It seems a remarkable coincidence, at least, that where dioxin and the compounds with which it is associated, even at supposedly minute levels, are concerned, the further one looks, the more trouble one finds.

Dr. Waldvogel insisted that the production of trichlorophenol at the ICMESA plant prior to the July explosion had been carried out with conscientious regard for the extremely toxic nature of dioxin insofar as the well-being of both the ICMESA workers and the public in the vicinity was concerned. He told me also that the dioxin content of the trichlorophenol product was regularly monitored through analytic techniques. This appears to be true, for samples of the trichlorophenol were regularly sent to Givaudan's research headquarters in Switzerland for analysis; after all, the company researchers certainly knew that if dioxin were to appear in Givaudan's hexachlorophene in large amounts, the results could prove disastrous to those on whom the preparation was used. When I called upon Dr. Peter Schudel, vice-director of research for Givaudan, in Dubendorf, he spoke with pride of the precautions taken at the Swiss laboratories in the handling of dioxin: only male technicians were employed in the analytic work; they worked in an isolated building and were regularly subjected to intensive medical examinations.

But at Givaudan's Italian plant the safety standards have been another story. The workers there were given no meticulous medical checkups such as were given workers in Givaudan's Swiss labs, and from what they have been reported as telling their union representatives and the press, not only were they not instructed by the ICMESA management concerning the true toxic properties of dioxin but they had no idea of its existence. As for the dioxin-laden residues that were a regular by-product of the ICMESA reactor, what became of them? According to the Givaudan management, the toxic waste was burned in an incinerator on the plant premises in such a way that the dioxin contaminant was destroyed. There is no indication, however, that emissions from this process were regularly monitored for dioxin content. (And, as I discovered in the course of my research, the supposedly destructive effects of burning on dioxin, except under the strictest of conditions, are highly debatable.) There was no way of telling what amounts of dioxin might be regularly escaping from the ICMESA plant into the air and over neighboring communities. From time to time during the two years in which ICMESA was producing trichlorophenol, nearby farmers would bring in dead rabbits—which they had been raising for food—and complain that pollution from ICMESA was responsible for the animals' deaths. The ICMESA management—"knowing that it's Italy," Dr. Waldvogel said—would pay them a small sum to settle the matter. As for Italian governmental or provincial regulations regarding the disposal of complex toxic by-products, for practical purposes there simply were none. Industry in Italy, much of it in the north under the control of foreign capital, operates almost entirely free of such impediments. It is perhaps characteristic of the chaotic situation prevailing that although the importation of 2,4,5-T into Italy was supposedly halted in 1970—following the impact abroad of the revelations at the Hart subcommittee hearings and of the decision of the American military to suspend the herbicide's use in Vietnam—the production of the very dioxin contaminant responsible for the original ban on 2,4,5-T could proceed at ICMESA without the slightest check. "We're a neocolonial society here," one of the Italian epidemiologists studying the aftereffects of the Seveso explosion remarked to me. I asked Dr. Waldvogel why neither ICMESA nor Givaudan appeared to have informed the authorities or the local people immediately after the ICMESA explosion that the chemical cloud from the factory might contain dioxin, and Dr. Waldvogel replied that representatives of the two companies *had* spoken about the possibility that "highly toxic by-products" were contained in the cloud. "If we had used the word 'dioxin' in consulting, for example, the Mayor of Seveso, would that have changed the situation?" he asked.

The amount of dioxin that the population of the area in and around Seveso was exposed to as a result of the explosion is a matter of conjecture. Estimates of the amount that I heard during my visit to Milan ranged from five hundred grams (about one pound) to five kilograms (or about eleven pounds). How much actually was released obviously will determine the extent of the effects on the population nearby, and, because dioxin poisoning appears to work not only in an immediate, devastating way but in many ways that are subtle and not properly understood, only time is likely to reveal some of the less manifest effects.

As it happens, a fair proportion of the burnlike skin lesions that many inhabitants suffered, which press reports assumed were symptomatic of chloracne, generally receded over the months following the explosion. These lesions were probably a result of direct contact with the sodium-hydroxide and phenolic components of the fallout. But, starting in late September, two and a half months after the cloud descended on the area, an increasing number of children and young people from Zone A began to develop on their faces, arms, and bodies symptoms of true chlor-acne—the sure mark of dioxin poisoning. By the beginning of November, forty-four people had developed confirmed cases of chloracne, and by the beginning of December the number had risen to forty-eight. A year after the explosion, the confirmed cases of chloracne were to rise to more than three times that number. There were other characteristic symptoms besides the skin acne, as I learned when, in October, 1976, I talked with members of one affected family from Seveso who had been eating in the yard outside their house when the toxic cloud enveloped them. The parents had washed themselves immediately, but their two daughters, who were about twelve and fifteen years old, had not; and when the cloud had apparently dissipated the family had continued with their outdoor meal, which was now coated with a film of dioxin-contaminated residues. All four of them became nauseated and suffered bouts of diarrhea. When I saw the family, both children had developed chloracne. Their faces were covered with an angry-looking rash. Their mother told me, out of their hearing, that one of the most distressing aspects of the episode had occurred even before the appearance of the chloracne, and this was what seemed to be a "complete change of character" in the two girls. In the past, they had possessed sunny, easy-going temperaments, she said, but now they had become extremely nervous and moody, with frequent irrational bursts of irritability. They had also suffered a marked loss of appetite. She told me that the same symptoms were prevalent among many of the people from Zone A, among those without chloracne symptoms as well as among those with chloracne. The mood of the men, especially, was affected, she said; they were strangely different—"very, very nervous," moody, irritable, and, at the same time, extraordinarily tired. On hearing this, I asked the woman's husband whether he had had any intimation that some men from Zone A had a reduced sexual drive—something that was associated with dioxin poisoning in previous industrial accidents. He said there was nothing to the notion. At that, his wife exclaimed, "The men don't want to speak of it, but all the women say it's true!" Though I was aware of the depressing effects of the toxic cloud on the ordinary life and good spirits of the people in Seveso, I could not help wondering whether the more striking personality changes I had been hearing about among the refugees might not be providing a warning sign of dioxin poisoning that had not yet manifested itself physically in the outward form of chloracne.

As for possible evidence of any teratogenic effects upon pregnant women in the area, the epidemiological picture has been blurred by the fact that the pregnant women who were permitted to receive therapeutic abortions—and those who obtained illegal abortions—were those who were reckoned to have been most heavily exposed to dioxin immediately following the ICMESA explosion. (These included women who lived outside Zones A and B.) Altogether, the legal and illegal abortions probably total more than a hundred and fifty.

A survey by an epidemiological commission appointed to study the health of people in the broader area of Meda, Seveso, Desio, and Cesano Maderno has shown that a hundred and eighty-three babies were delivered in the two months following the ICMESA accident and that there were fifty-one spontaneous, as distinct from induced, abortions. According to these figures, the percentage of spontaneous abortions in relation to pregnancies would thus be twenty-two per cent, which would be approximately double the rate previously recorded for the area. (But it seems fair to add that the accuracy of such a comparison may be affected by the degree to which Italian women receiving illicit abortions may have managed, with the help of friendly physicians, to have these abortions registered as having taken place spontaneously and involuntarily.)

At the time of my visit to Milan, some eight cases of birth abnormalities among babies born to women in the Seveso area who were pregnant at the time of the ICMESA explosion had been reported. But in the opinion of Professor Gaetano Fara, the chairman of the epidemiological study team appointed by the Lombardy regional authorities to examine the effects of the ICMESA explosion on the health of the local population, any cause-and-effect relationship between the toxic cloud and these abnormal births is difficult to infer, because, Dr. Fara told me, the incidence of these birth abnormalities is not really disproportionate to the incidence of those that might normally be expected to occur in the locality.

However, it would be unrealistic to conclude that the babies of local mothers pregnant at the time of the explosion are necessarily to be considered free of untoward effects of any exposure to dioxin that their mothers may have suffered. Experience with previous accidents involving dioxin indicates that the systemic effects of the contaminant are generally delayed and appear over long periods of time, and, furthermore, are hard to predict or to measure accurately. For example, who could pinpoint a cause-and-effect relationship involving dioxin if a child of a dioxin-exposed mother should turn out to be just a little bit less alert than the average, a little more prone to memory lapses, a little more "nervous," or a little more susceptible to illnesses of various kinds? Virtually no research has ever been carried out on the children of dioxin-exposed mothers whose husbands were involved in accidents in 2,4,5-T plants. In fact, so little is known about the full, long-term consequences of human exposure to dioxin that it is safe to say that there exists not one scientist anywhere in the world who could truly be called an expert on the effects on people of this tremendously toxic substance.

IN the meantime, Zone A at Seveso remains a desolate and barren area, peopled only by the masked, white-suited figures taking soil samples for analysis or, now, engaging in efforts to prevent contaminated material within the area from drifting or being blown outside it. The dioxin contaminant in the soil within Zone A is not just on the surface but persists in the topsoil to a depth of about twelve inches. The degree of contamination varies considerably, and the variation between Zones A and B is even greater. In the soil of certain parts of Zone A, the amount of dioxin has been measured in milligrams per square metre—a very heavy amount indeed. The total amount of the contaminant in Zone A has been estimated at between one and eleven pounds. For Zone B, the estimates are very much lower, running at less than an ounce—a figure that, considering the toxicity of dioxin, is still a very significant one. The problem of how, if ever, Zone A can be sufficiently decontaminated to make it reasonably safe for people to live in is one that the Italian authorities are still grappling with, and to which they obviously have no ready answers. The plans now being considered for all topsoil whose dioxin contamination is higher than five micrograms per square metre include physical removal by bulldozer, chemical and biological treatment, and high-temperature incineration. This involves the effective disposal of up to three hundred thousand tons of topsoil, a task that would require six months simply for the bulldozing stage and a minimum of two years for some sort of incineration process.

Some scientists have expressed a belief that most of the dioxin in the soil can eventually be destroyed by a combination of photodegradation (by the action of sunlight) and biodegradation (by naturally occurring soil organisms and by bacteria specially selected and propagated so as to have the capacity for attacking and breaking down dioxin molecules). Givaudan claims to have had success in reducing the amount of dioxin residue on vegetation in a selected experimental plot within Zone A by spraying the vegetation with an olive-oil solution and the chemical cyclohexanone and thus creating chemical interactions that speed up the process of photodegradation. However, Dr. Vittorio Carreri, who heads the decontamination study team appointed by the Lombardy provincial authorities, has called the results of the Givaudan experiment "poor." A recent United States Air Force study of residues of Agent Orange on experimental plots in Utah, Kansas, and Florida, while it found the dioxin contaminant of Agent Orange to be relatively persistent, also concluded that the contaminant could be very substantially reduced over a period of time by the action of sunlight. Yet the sanguine forecasts of some scientists concerning the degradability of dioxin and the feasibility of decontaminating Zone A to the point of making the area a safe place to live in are hardly borne out by the history of decontamination attempts that have been made in the aftermaths of previous dioxin-producing industrial accidents. Accounts of these attempts indicate that dioxin can persist and continue to have poisonous effects for a considerable period not only after the initial accident but also after the most elaborate decontamination measures have been undertaken.

An incident following the explosion in the Badischer Anilin & Soda-Fabrik's 2,4,5-T factory in Germany in 1953 illustrates the capacity of dioxin for lingering in the environment and continuing to exert ill effects. Five years after the accident, an employee was assigned to perform some work on or near a reactor vessel that since the original accident had not been used. For the task, the worker wore a completely protective suit, including a face mask. However, in the course of the work, he removed the mask several times to wipe perspiration from his face. Within four days, he developed headaches, loss of hearing, and chloracne. A month later, he was hospitalized with angina pectoris. Six months after that, he developed pancreatitis and a tumor in the upper abdomen. He died some nine months after his exposure to dioxin. Another workman, who, also five years after the B.A.S.F. explosion, spent only two hours repairing an interior wall in the area containing a reactor vessel, likewise developed severe chloracne. A year later, a large X-ray-opaque area developed in his left lung. Four years after that, he suffered acute psychosis, with the hearing of voices, and committed suicide.

Another example of the seemingly random manner in

which dioxin can exert ill effects after long periods and after intensive efforts have been made to remove all measurable traces of it, is to be found in an incident in the aftermath of the Coalite trichlorophenol factory explosion in Bolsover, England, in 1968. The only factory equipment that was not buried deep underground after that explosion was the several large tanks that had been rigorously and repeatedly cleaned with high-pressure steam jets to remove any possible traces of dioxin. Nearly three years after the explosion, two pipe fitters were assigned to connect new equipment to new collars that other workers, some time previously, had fitted onto one of these steam-cleaned tanks. Within four weeks of doing this work, both men developed severe chloracne. Within eleven months of the date of exposure, the wife of one of the workers developed symptoms of chloracne, as did the son of the other worker, within three months. Yet subsequently careful analysis of the surface of the tank and the fittings failed to reveal any traces of dioxin.

Still another example of the persistence of dioxin contamination is the experience of a group of eighteen workers who, six months after the 1963 explosion in the Philips Duphar 2,4,5-T factory in Amsterdam, were put to work in an attempt to decontaminate the plant. All worked in deep-sea diving suits and—except for one man—in goggles and industrial face masks. Despite these precautions, nine of the workers subsequently developed chloracne, and, of these nine, three died within two years. The one man who did not wear a face mask or goggles was reported to be severely affected and unable to work; it was reported that he was still requiring treatment in 1976. (It may have been partly on the basis of this experience that the Philips plant was eventually dismantled and dumped at sea.)

The proposals for breaking down the dioxin contaminants in the Seveso area through high-temperature incineration have been received with skepticism by some scientists. The problem is that the thermal, mechanical, and other conditions necessary to assure the complete destruction of dioxin on any large scale must be carried out entirely without error, if further disaster is to be avoided. One American scientist who is familiar with all such difficulties, and whom I met in October, 1976, in Milan while he was consulting with the Italian authorities on the question of dioxin contamination, declared that he considered the proposal to incinerate dioxin-contaminated material from Zone A on the spot "a dangerous course" and "the most risky thing in the world." The evacuees from Seveso, without possessing any particular scientific knowledge, are in strong agreement with this view. Since the last months of 1976, the opposition of many of the Seveso inhabitants has become so vehement that the authorities have grown dubious about the political practicability of getting an incinerator built in the affected area. And this situation presents them with a whole new set of problems unpleasant to contemplate. If they have to erect an incinerator in some other part of Italy to handle the polluted soil of Seveso, what place could they decide upon? And if they do find a new incineration site, how could they cope with the dangers of transporting to it hundreds of thousands of tons of dioxin-contaminated soil? As an alternative, they could, of course, bury all the contaminated material deep in the earth, but, if so, where? Under already polluted Seveso? And if not under Seveso, where? And, again, how could it be transported there?

In the meantime, the Italian authorities are still trying to deal with manifold problems. One of the serious ones they faced in the autumn of 1976 was how to contend with the fall of dioxin-contaminated leaves from the trees, since there was every likelihood that winter winds could blow this poisoned foliage into so far unpolluted areas. Furthermore, other vegetation in the area was going to seed, so there was the additional danger of clouds of fine particles of vegetable matter and chaff being blown around, and being inhaled by and otherwise coming into contact with people outside the evacuated zone. The authorities sent tractor teams manned by safety-suited workers into the area to cut down and mechanically bale vegetation in the larger open areas. In smaller areas, the cutting was done with mechanical sickle bars or old-fashioned scythes; no rotary mowers were permitted, because of the hazards posed by chopped-up cuttings flying around. As for the autumn foliage, the authorities had all limbs lopped from the deciduous trees, including fruit trees. As the limbs were removed, the leaves on them were picked off by hand and put into large plastic bags for removal to a designated storage area in the zone, where all this material will be kept until some method of ultimate disposal can be decided upon. The limbs themselves, after being lopped and sawed into lengths, but before being carted away to the disposal area, were sprayed with a foamlike solution of polyvinyl acetate to help keep the bark from crumbling off. Despite all these efforts, a large amount of dioxin-contaminated leaf debris continues to lie and blow around in Zone A, presenting yet another source of continuing concern. With time, it has become clear that dioxin contamination exists well beyond the area officially designated as polluted.

The work of the scientific groups appointed in Italy to study the effects of the ICMESA explosion has been conducted in such an atmosphere of confusion, and the issue of the possible effects of dioxin contamination on the unborn has become so entangled with considerations of religious doctrine and Italian politics (and even further complicated by scientific factionalism), that whatever conclusions these investigators may draw are almost certain to be the subject of sharp controversy. Given the contentious atmosphere in which the present and future lot of the people of Seveso is being assessed, one is led to give particular attention to the lessons that may be drawn from the measurable effects of dioxin in controlled experiments using laboratory animals. Two recent studies by a team led by Dr. James R. Allen, professor of pathology at the University of Wisconsin Medical School, appear instructive. The first study concerned the long-term effect of dioxin-contaminated food on non-human primates—specifically, on eight female rhesus monkeys.* For nine months, the monkeys were fed a diet containing dioxin in a concentration of five hundred parts per trillion. Within six months after they received this diet, all the monkeys developed anemia, and within nine months the survivors also suffered severe reductions in red and white blood cells and platelets. Between the seventh and twelfth month, five of the eight animals died. The results of this study, taken together with previous animal studies conducted by Dr. Allen and his associates, appear to demonstrate impressively the ability of dioxin to persist and accumulate in the living tissues of primates.

In a second, eighteen-month feeding study, conducted on seven groups of experimental rats,† the findings not only fully confirmed the extreme and accumulative effects of dioxin in terms of observed liver and kidney damage and anemia but pointed to an even more serious effect. Out of sixty animals under study, twenty-three developed tumors.

---

*See Appendix, p. 168.

†See Appendix, p. 178.

Examination showed a significant increase in tumors in a group of rats receiving daily dioxin levels as low as five parts per trillion in the food. "The high incidence of neoplasms in rats fed subacute levels of TCDD suggests the carcinogenic potential of the compound," a scientific paper written by Dr. Allen on the study has noted.

This, then, is the nature of the chemical that has left its poisonous and enduring trail in so many parts of the world, and remains as an inevitable, even if presently reduced, contaminant of the 2,4,5-T herbicide that was rained down by the American military in such vast quantities on Southeast Asia and is now being rained down, year after year, in even vaster quantities, over millions of acres within the United States—even as the military itself has at last disposed of its almost two and a half million gallons of Agent Orange. This is the toxin whose production and use responsible officials of the E.P.A. seem powerless or unwilling to forbid or effectively control. Not only has E.P.A. become virtually paralyzed where control of the 2,4,5-T problem is concerned but most of the vast 2,4,5-T herbicidal-spray programs have actually been carried out and continue to go forward under the auspices and at the instigation of other federal agencies—in particular, of the United States Forest Service and the United States Department of Agriculture. In the light of what is now known about the almost unmatched toxicity of dioxin—and of what is *not* known about its precise mode of action and its long-term effects on human beings—should the government accept the assurances of herbicide manufacturers that a little dioxin is "acceptable" to people, just because the dioxin contaminant is being spread and sprayed around over large, rather than small, areas?

The government itself appears to have accepted the assurances of the Dow Chemical people and other herbicide manufacturers that the dioxin contaminant in 2,4,5-T, once laid down, is degraded and effectively destroyed by the action of sunlight and of soil bacteria. Yet if the contaminant is in fact so handily decomposed in this manner, one wonders how to account for the results of an analysis of samples taken in 1975 of the fatty tissues of cattle that grazed on Western rangeland that had been sprayed with 2,4,5-T during the previous year, for the samples showed significant levels of dioxin—levels of up to sixty parts per trillion. How is it that from grazing land sprayed with this allegedly innocent herbicide the government permits the distribution to the American dinner table of meat that has been contaminated with measurable amounts of one of the most toxic substances known to man? It is not as though the government were unaware of such findings, or of their significance, for the very studies that produced these findings were conducted under the auspices of the E.P.A. And a memorandum dated August 5, 1975, gives these observations of the dioxin-study program coördinator in the E.P.A. Office of Pesticide Programs concerning the significance of the findings:

> Studies including teratogenic and other toxicity effects indicate that the residue levels mentioned above may present a health hazard to man based on the application of normal margins of safety.

Almost incredibly, the E.P.A. people still cannot make up their minds. The issue is completely stalled, and action on it is blocked. In the meantime, Dr. Patrick O'Keefe and Dr. Matthew Meselson at Harvard have obtained several samples of mother's milk from Texas and Oregon, and subjected them to analysis for possible dioxin content. The preliminary results of this latest study indicate the presence of dioxin in parts-per-trillion amounts in some of

3

RETURN TO SEVESO SEPTEMBER, 1978

JUST BEFORE the beginning of 1978, I decided to return to Seveso to inquire into the condition of the people there nearly a year and a half after the toxic cloud released by the ICMESA explosion had descended on the town and the surrounding area. Since my last visit, not much information had been reaching the United States concerning the situation in Seveso. Media interest in the aftereffects of the disaster had declined. Most of the news items I came across on the subject during this period were from the English press, which seems to have made more of an attempt to keep abreast of developments in Seveso than American newspapers did. One item concerned the discovery of some unexpectedly elevated levels of dioxin and some suspected cases of chloracne that had been detected outside the boundaries of Zones A and B; another reported on more demonstrations by local citizens in opposition to construction of the huge incinerator that had been proposed as a means of destroying the hundreds of thousands of tons of dioxin-contaminated material within Zone A; other items noted that the regional authorities were so satisfied with progress in decontaminating considerable parts of Zone A that they were preparing to return many of the people to their homes within those parts of the zone so that the people could resume something like a normal life.

When I arrived in Italy to learn more about the situation in Seveso, I was relieved to learn from a toxicologist in

Milan whom I had interviewed in 1976 that, with the principal exception of those still afflicted with chloracne, most of the people who had been evacuated from Zone A had survived their exposure to dioxin better than had been expected. I was particularly relieved to hear that although in 1977 several babies in the area had been born with malformations, early fears that birth deformities would occur on a wide scale among the children born to women in the area had so far not been realized, and that the incidence of malformations that had occurred did not seem to be beyond the range of what might normally be expected in the population.

The next day, I went to Seveso and called on Mayor Rocca at his home. He lives in a part of town that is on a hillside that rises above an otherwise flat terrain. Here winding roads are lined with substantial villas within walled enclosures. Obviously, these are the homes of many of the Seveso élite—lawyers, company executives, timber merchants. I noticed that the outer walls of several of the properties had been spray-painted with swastikas—evidently the work of leftist protesters. The Mayor's house bore no such signs. It is a relatively new two-story red brick house of considerable size. Mayor Rocca, a man with short curly hair and a sharp, almost upturning Punchinello chin, received me graciously in his living room. I reminded the Mayor that I had met him on my previous visit, and that he then expressed considerable anxiety about the uncertainties facing the people of Seveso after the ICMESA explosion, and I told him I wanted to learn about the present welfare of the community and what the mood of the people was almost two years later.

Mayor Rocca told me that he was pleased to be able to say that of the seven hundred and thirty-nine people who had eventually had to be evacuated from Zone A—which had been somewhat altered by the Lombardy regional authorities some time after the explosion—more than a hundred and fifty were moving back into their houses in the lower part of the zone, which was now considered free enough of dioxin to be safe for habitation. Zone A, he explained, had been divided into seven subzones, numbered according to their distance from the ICMESA plant, and the part being reoccupied consisted of Subzones A-6 and A-7. All this aroused thankful feelings among the affected people, not only in Subzones A-6 and A-7 but in the entire surrounding area as well, the Mayor said. "I am glad to inform you that at the end of this week there will be a major event," he continued solemnly. "You should know that Giovanni Cardinal Colombo, the Archbishop of Milan, promised a year ago, at a special service he preached in the unevacuated part of Seveso, that he would come back to Seveso when the people returned to their homes. And now, this Saturday, sure enough, he will come to Subzones A-6 and A-7 to bless the houses that the people are reoccupying. The Archbishop's blessing gives, so to speak, the official seal of approval to the return of the people to their homes. It sets the tone. There is a strong feeling among the people that, while nobody is likely to forget what has happened here, for them the worst is over. They believe that their trials are behind them." The Mayor added, "The authorities feel that the measures they took right away—the evacuation and so on—were prudent and prevented major health disasters. The fact is that none of the strange diseases that were predicted everywhere for the Seveso people have been detected."

On the abortion issue that arose after the accident, the Mayor left no doubt about where he stood. "In a surrounding area of a hundred thousand people, including the people of Seveso, thirteen hundred children have been born without any of the malformations forecast," he told me. "There were certain political forces—radicals, left-wing parties—that took advantage of the accident and latched on to the abortion issue. They poisoned the atmosphere. But in fact the population did not take advantage of the opportunity to obtain abortions. These thirteen hundred normal children were born despite all the dire warnings. In August or September of 1976, a group came to Seveso demonstrating for abortion, trying to get the women up in arms. They failed totally. Most of the demonstrators were radicals. Some of the mothers in Seveso came to consult with me on the question. I am very happy that they had the courage to bring their children into the world. The local women chose life."

The Mayor said that through the influence of Carlo Cardinal Confalonieri, who is the Dean of the College of Cardinals, and who himself comes from Seveso, he had been admitted to an audience with Pope Paul VI in September of 1977 and had discussed the Seveso incident with His Holiness. "The Pope commented that he thought the Almighty had a hand here," Mayor Rocca said.

I asked the Mayor how he would sum up the impact of the toxic cloud and its consequences for the people.

He replied, "Seveso has been held up as an ecological disaster. The people don't feel that that is the reality. After all, in these past seventeen months they have lived through all this tension and managed to carry on their lives in spite of the uncertainties and risks. I think these things should be held up to mankind as an example of how a community hit by heavy misfortune could come through it."

In the days after this encounter, it seemed to me that the views expressed by Mayor Rocca—that the worst was over and the time was auspicious for a return to normal life in a dioxin-free environment—represented the official attitude toward the present condition and outlook of the people of Seveso. And it also appeared—on the surface, at least—to express the views of most of the evacuees I talked to who had already returned or were in the course of returning to their houses in the reopened sections of Zone A. The more I saw and heard, however, the more I became aware that the situation facing the people of Seveso and the surrounding area was far more complex than many of them cared to admit, certainly to outsiders, and perhaps even to themselves.

FROM the very beginning, the problems of coping with the threats to the health, safety, and welfare of the inhabitants of the area were greatly complicated by political and economic considerations. In addition, the affected people faced innumerable difficulties caused by bureaucratic delays, factional bickering, and, it appears, frequent bungling and mismanagement of the programs that the regional authorities had devised to cope with the poisoning of the area. All this, one gathers, played a part in instilling within at least a certain part of the population deepening suspicions of what the authorities might have in store for them, and over the last year such suspicions have resulted in a noticeable lack of cooperation with many of the official programs and sporadic bouts of active resistance to them.

A brief look at a map of Seveso and the surrounding area issued by the regional authorities which displays a revised demarcation of Zones A and B is enough to give one some indication of how public-health data concerning dioxin contamination has been a subject for political compromise or manipulation from the start. Within the cone-shaped area that was actually covered by the toxic cloud, the eastern boundary of the most severe contamination is depicted on this map as stopping in miraculous fashion along the straight western edge of the Milan-Como superstrada. It seems clear that this boundary was, in effect, gerrymandered by the regional authorities to exempt commercial and private traffic on the highway from the severe restrictions placed on all other traffic in the area, including traffic in the less contaminated Zone B. In excluding the superstrada from the sealed-off area, the authorities did not make known to the public who might use the highway what they themselves must surely have known—that some of the higher concentrations of dioxin in the entire area were to be found along the superstrada. (At one point in the superstrada system, at an exchange near the ICMESA factory, the concentration of dioxin is reported to have been measured at more than five thousand micrograms per square metre—a dangerously high level.) To judge from the events of subsequent months, the decision to allow a large flow of traffic to continue through the heavily contaminated stretch may have had unfortunate consequences for people living outside the immediately affected area.

Although it was only natural for those in charge to be wary of exacerbating the fears of the people or creating a condition of panic among them, it appears that from the outset the authorities showed a strong interest in limiting the zone marked as heavily contaminated to as small an area as possible, and in treating the question of the hazards to which the population of Zone B and the Zone of Respect was subjected as very much a secondary matter. The difference in the official attitude toward the hazards existing in Zone A and those existing in the two other zones, combined with all sorts of inconsistencies in treating the dioxin threat in and immediately around Zone A, seems to have created, in the minds of many of the exposed people, not only confusion and a sense of frustration but also a certain cynicism about the safety measures employed—a cynicism that may have affected the attitude toward the dangers of the dioxin itself.

Even before the evacuation of Zone A was completed, some of the people in Seveso began to question how strongly the Lombardy regional authorities believed in the threat of dioxin. For example, late in July, while people were still living in Zone A, the authorities first forbade all entry into it by outsiders and then changed the rules to allow ten persons to enter on weekdays and twenty on Sundays. "When people in the area heard about these safety measures, some of them just laughed, in spite of the tragedy they were caught up in," one person familiar with the evacuation told me. "They would say to one another, 'So dioxin is twice as dangerous on weekdays as it is on Sundays?' " People noted that access to Zone A was guarded by soldiers who were supplied with neither face masks nor protective clothing. They also noted that no restrictions whatever were being placed on traffic moving through the area on the superstrada. Those who were evacuated from Zone A, although they had been formally notified that they would not be allowed to take with them anything but their clothes and some personal belongings, took as many of their possessions with them as they could manage. These possessions included furniture, cars, pickup trucks, baby carriages, and bicycles. Inevitably, many of the things taken out of Zone A were contaminated with dioxin.

Human nature being what it is, some of the people who had continued to eat fruit from their trees in spite of official warnings that to do so might be hazardous seem to have chosen to view dioxin as something less than a serious threat to health. Such a view was reinforced by doubts publicly expressed from within the Christian Democratic establishment concerning the gravity of the dioxin problem. On August 2nd, when the second group of people was evacuated from Zone A, one regional health official was quoted as assuring the public that "to die of dioxin, it would be necessary for a child to eat three hundred kilos of earth" from Zone A. Dr. Emilio Trabucchi, professor of pharmacology at the University of Milan School of Medicine and a member of the Italian Chamber of Deputies, publicly stated that the dioxin presented no real threat to health, and, to demonstrate that he meant it, Professor Trabucchi declared himself willing to go and live in Zone A and to drink dioxin-contaminated milk produced by cows on a small farm near the ICMESA factory. As it happened, though, the cows in the zone had to be slaughtered to protect public health.

The general effect of such declarations was, of course, to add to the confusion and to make people reluctant to accept the pronouncements of the regional health authorities—reluctant, even, to accept the good faith of those charged with helping to cope with their difficulties. The evacuees soon became vocal about the failure of the authorities to find them new houses to live in, rather than rooms in the hotel complex in Bruzzano. At the same time, there was growing discontent among them over the failure of the authorities to take some definite steps toward making their homes in Zone A habitable once more. Early in October, after the Army troops guarding access to Zone A had been largely replaced by carabinieri, several hundred of the Seveso evacuees broke through barbed-wire barriers and past police at the entrance points and reoccupied their old homes. "The demonstrators said they felt the danger from [dioxin] would have been washed away by recent heavy rains," a dispatch in the London *Times* reported. After holding out in their homes for a day, these people were persuaded by police to leave the zone. In twos and threes, or in larger groups, they came back, however, and, the Italian sense of home and family being what it is, the carabinieri guarding access to the zone were apparently lenient about letting people reenter.

Eventually, in an attempt to tighten security, the regional authorities ordered a fence about nine feet high and four miles long erected around the entire zone. It took the authorities seven months to complete the fence. The extraordinarily long time taken to finish the job was due partly to inefficiency but partly also to the fact that in places the fence was torn down at night as fast as it could be put up by workmen during the day. Through these breaches in the fence evacuees kept up a regular traffic back to their houses. Most went in to check up on whether their houses remained undisturbed—some vandalism and looting within Zone A had been reported—but people were also removing their furniture to their new quarters outside Zone A. Some of the abandoned houses in the most contaminated zone were said to be used for lovers' assignations, and it was also said that on weekends children from Zone B had been seen playing and roaming around inside Zone A, where, presumably, they felt that their parents wouldn't be keeping an eye on them. In the third week in February of 1977, an Italian Army contingent of five hundred men of the 67th Mechanized Infantry Battalion was called in at the request of the regional authorities to protect the fence and prevent further unauthorized entry into the zone. But by that time a great deal of damage had been done.

An inevitable result of all the traffic in and out of Zone A was the spread of dioxin—by particles borne on clothes, furniture, and bedding—from the most contaminated zone into areas officially classified as habitable. Actually, this situation serves as a telling commentary on the contention maintained for years by the Dow Chemical people in this country. In reply to charges concerning the spread of the dioxin contaminant of 2,4,5-T into the environment here from civilian aerial-spray operations being carried on around the country, the Dow people have said that dioxin residues, once laid down in spray or other operations, do not at all readily move around in the environment—first, because the dioxin becomes chemically bonded to soil it touches, and, second, because it is for all practical purposes insoluble in water and thus is not susceptible, for example, to leaching action. The aftermath of the Seveso accident shows that a number of forces, some of them natural and others the work of man, created conditions for the transport of dioxin-contaminated matter from the most polluted zone outward. These forces included strong winds that could blow soil particles and contaminated leaves, chaff, and other vegetation out of the zone. They also included the heavy automobile traffic sweeping along the edge of Zone A—traffic that was picking up dioxin contamination and carrying it north and south of Zone A and the whole Seveso area. And they included persistent rains. In October of 1976, unusually violent and continuous rains afflicted the whole of northern Italy for more than a month and caused extensive flooding of the most polluted part of Seveso. These floodwaters, bearing dioxin-contaminated soil particles, ran off into the Certesa, a stream that flows past the ICMESA plant and down to the eastern edge of the Zone of Respect, where it joins the Seveso River. During these fall floods, both the Certesa and the Seveso overflowed their banks. The Seveso runs through several communities to the south, including Cesano Maderno, and to the city limits of Milan itself, where it disappears into underground conduits and emerges to the south of the city. There the Seveso joins the Lambro, which joins the Po. Traces of dioxin have been found in mud as far south as Milan, thirteen miles away. Also, significant dioxin levels—

whether spread directly by the toxic cloud or indirectly by the Seveso inhabitants themselves—kept turning up outside Zones A and B for months after the evacuation of the residents. And the incidence of chloracne among children of the inhabitants continued to increase.

By the beginning of November of 1976, as I have previously noted, out of several hundred suspected cases the number of confirmed cases of chloracne in the area was forty-four. Eventually, the number of confirmed cases rose to a hundred and eighty-seven. In September of 1976, the authorities had declared all schools attended by children in the area to be free of all traces of dioxin, but by early February so many children had come down with suspected cases of chloracne in six schools attended by children outside Zone B but inside the Zone of Respect that mothers were refusing to allow their children to attend them. As a result, the schools that were found to be contaminated were closed by the authorities in early February so that they could be decontaminated. According to the health authorities, epidemiologists suspected that schoolchildren had tracked dioxin into classrooms on the soles of their shoes. If so, some of the dioxin contamination may well have had its origin in the constant unauthorized traffic of Seveso families between their new quarters in the supposedly unaffected zones and their old homes in Zone A.

Significant signs of dioxin contamination were also turning up in other areas outside Zones A and B. In early 1977, it was discovered that in Polo, a community of thirty-five hundred people, lying to the east of the ICMESA plant, which had been considered only slightly contaminated and made part of the Zone of Respect, fifty-eight out of two hundred and eighty-three children examined had developed symptoms that were tentatively diagnosed as similar to those of chloracne. (Because ordinary facial acne is common among young people, the number of these children subsequently found to have true chloracne was bound to be much smaller than the number of suspected cases. But at least a dozen of the Polo children probably contracted true chloracne.) Then, in April, it was discovered that in the township of Cesano Maderno the grounds of fifteen small factories employing a hundred and eighteen workers were contaminated with dioxin at levels comparable to those in some parts of Zone A. According to Vittorio Rivolta, the head of the Lombardy regional health organization, these particular instances of contamination were thought to have been caused primarily by vehicles coming to the factories from more contaminated areas. After closing the factories briefly to check on dioxin levels, the authorities decided to deal with the situation by reopening the factories and ordering that the grounds be covered with asphalt.

In the meantime, the measures being taken or planned by the Lombardy authorities in the hope of reducing or eliminating dioxin contamination in the area were themselves presenting serious problems. Since the three sections of Zone A closest to the ICMESA factory were considered so contaminated as to be unreclaimable and uninhabitable within the foreseeable future, most of the organized efforts aimed at making it possible for evacuees to return to their homes were concentrated on Subzones A-6 and A-7, in the southern part of Zone A. There the initial reclamation plans called for the bulldozing and removal of all soil contaminated with dioxin at levels of above five micrograms per square metre. But this was an arbitrary figure, chosen in the rather forlorn hope that lower levels of dioxin in the soil would degrade naturally to insignificant levels in a reasonably short time—perhaps two or three years. This standard meant the bulldozing and removal of many thousands of tons of topsoil from Subzones A-6 and A-7 alone and the importation of uncontaminated topsoil from other parts of northern Italy to replace it.

Before the end of 1976, workmen began bulldozing contaminated earth into large piles; then the earth was loaded onto trucks and dumped in the more heavily contaminated part of Zone A. This was the part of Zone A that the authorities had decided on as the site for the high-temperature incinerator where the contaminated material would be burned to destroy the dioxin in it. (The soil itself would necessarily be destroyed as well.) Workmen also began attempting to decontaminate the houses in A-6 and A-7 by replacing roof tiles, washing and scrubbing down floors and walls, and removing many of the furnishings left in the houses. These contaminated materials were then trucked to dumping points in the more heavily contaminated part of Zone A. Chopped-down bushes, limbs of felled trees, and other vegetation from A-6 and A-7 were also added to the large piles within the more highly contaminated parts of Zone A. And then there were the piles of plastic bags containing the decayed carcasses of the thousands of animals and birds that had either been killed by dioxin during the first several days after the explosion or been subsequently slaughtered for public-health reasons. All this material was supposed to be incinerated at some future date.

The notion of incinerating such a vast amount of contaminated material in a huge furnace built within the township boundaries was, not unexpectedly, a disturbing one not only to the Seveso refugees but to the people still living within the unevacuated parts of Seveso and the surrounding communities. The principal objections had to do with the safety of the proposed incinerator. The thought of its operation gave rise in the minds of some of the people not to a vision of purified surroundings but to the spectre of some new and conceivably even more toxic cloud that, through inadvertence, might descend on them in the future. And there were other objections that were put forward at least as strongly.

THE effects of the toxic cloud on the local economy had been most serious. A great deal of the furniture produced in Italy originated in the Seveso area, and with all the publicity about dioxin contamination the sales of furniture from the area north of Milan dropped drastically. Before the explosion, newly married couples would often journey all the way from southern Italy to showrooms in the Seveso area to pick out bedroom and living-room suites, which they could buy at prices below the rates generally prevailing in Italy. Now they avoided the area and bought furniture originating elsewhere. Few outsiders wanted to take the chance of living with furniture that might be impregnated with dioxin. Some consignments of furniture for export were reported to have been turned back at the Swiss border. Seveso craftsmen considered this a particularly cruel touch, since the ICMESA factory was Swiss-owned.

The business of imported-timber merchants in the area was also seriously affected: sawmill owners were unwilling to handle stocks of stored timber possibly tainted with dioxin. Owners of and speculators in real estate found that there were no buyers for the properties. Hitherto prosperous masons, plasterers, and carpenters could get no new business. Contractors of all kinds were without work. Bakers found that many people would not buy bread, because they thought the stocks of flour might have been contaminated. The same applied to merchants who normally sold locally grown produce to people outside as well as inside the affected area; most of them either were without work or could get very few orders—a situation they were finding it extremely difficult to change. Even their contact with potential customers was being frustrated, because by order of the authorities access roads from the Milan-Como superstrada were closed to outside traffic. The fact that refugees from Zone A and some of the affected people from Zone B were being paid modest allowances by the Lombardy authorities (who were in litigation with the ICMESA company and its Swiss owners for the huge damages being suffered by the Seveso people) did little to alleviate the economic depression that had descended on the area along with the toxic cloud.

The prospect that these economic conditions might be worsened by uncertainties arising out of the plan to build a dioxin-burning incinerator in their midst caused some local businessmen at the beginning of 1977 to start organizing the growing opposition to the incinerator project. In December of 1976, a group of these businessmen, most of them owners of real estate, had led a demonstration of Seveso citizens who formed a roadblock across the Milan-Como superstrada, disrupting all traffic for a full day, to protest the continued closing of access roads to parts of Zone B from the superstrada and certain other roads. After this protest, the businessmen extended their campaign to embrace the issue of the incinerator, on the grounds of both public safety and the prospect of greater economic losses to the area. By this time, the regional authorities had decided to go ahead with plans for the incinerator in the central part of Zone A, and the municipal council in Seveso had passed a resolution agreeing in principle to the scheme. But the opposition of the local businessmen quickly spread to include many Seveso craftsmen, and in May an angry demonstration attended by several thousand people from the Seveso area was held in central Milan to protest the proposed incinerator.

The turmoil and protest that spring also included an act of terrorism. Late in the afternoon of May 19th, while Dr. Giuseppe Ghetti, the local health officer for Seveso, Cesano Maderno, and six other small townships in the area—at the time of the ICMESA explosion he had had the responsibility of overseeing industrial-health measures to protect factory workers—was at his office in Seveso, three armed men burst in. One of the intruders pointed a pistol at Ghetti's forehead and demanded documents relating to ICMESA and other corporations, and when Ghetti vigorously reacted by attempting to throw his assailant out of his office he was shot and seriously wounded in the legs—the form of assault against political victims made notorious by Italy's leftist Red Brigades. Subsequently, copies of a proclamation issued on behalf of the terrorists who shot Dr. Ghetti were found about the Seveso area. During my visit, I saw a copy, which bore the name of no issuing organization other than Fighters for Communism. It began with the salutation "Worker Companions!," and it stated that Ghetti had "been hit by an armed proletarian nucleus" at his Seveso office. The proclamation charged Ghetti with various "criminal activities," including failure to take action before the ICMESA explosion to protect scores of workers at the Cesano Maderno plant of a company called ACNA, which used aromatic amines in the processing of pigments and dyes, from the allegedly fatal poisonous effects—principally bladder tumors—of the factory's production process. Ghetti, the proclamation charged, was guilty of equally criminal behavior in failing to protect the health of the workers at the ICMESA plant and of the people in the surrounding area. It declared, "A man in the pay of the most criminal masters . . . G. Ghetti said nothing about the work being done at ICMESA. . . . Not only did he not say anything about [the real nature of the toxic cloud] after the July explosion, he hid and minimized what he had previously known and what today has become a spreading poison."

The attack on Ghetti, who is back at his job but still recovering from his injuries, was not the only terrorist act to be committed in connection with the Seveso disaster. On July 10, 1977, the first anniversary of the ICMESA explosion, terrorists bombed the home of Dr. Rudolf Rupp, an executive of the Hoffmann-La Roche Corporation who acted as liaison with the Lombardy authorities on decontamination procedures being carried out in Seveso, and who lived in Basel. Responsibility for the bombing was claimed by a group calling itself Commando July 10.

While these terrorist acts probably originated with extreme leftist militants operating out of Milan, rather than with people in the Seveso area, they certainly added to the anxiety of the Seveso people, and possibly increased the defensiveness of the regional authorities about their failure over the years to correct industrial practices and built-in hazards of the kind that had led to the ICMESA explosion. Also, the health authorities were becoming increasingly vulnerable to criticism because of their handling of programs to cope with the aftermath of the toxic cloud—in particular, their plans to contain existing contamination and monitor the health of people living in the affected area.

According to the authorities, the methods of bulldozing earth and cutting down vegetation in Subzones A-6 and A-7 and removing it to the heavily contaminated part of Zone A were being carried out efficiently and under rigorous safety procedures designed to keep decontamination workers from direct contact with dioxin particles and also to prevent any spread of dioxin from the work area into inhabited surroundings. All decontamination workers entering the fenced-off area were required to pass through a so-called filter station, where they stripped and changed into impermeable decontamination suits, boots, and hoods and put on face masks. After the work shift, they were required to shower and to hand in their decontamination suits, which were then placed in plastic containers for removal to contaminated-material dumps within the zone by the next work shift. The teams of workers assigned to bulldozing, vegetation cutting, and house decontamination in Subzones A-6 and A-7, which until late 1977 also lay within the fenced-off area, were supposed to be similarly masked and suited. Most of this work was carried out by a work force supplied under contract by a company in Milan. It appears that in practice many of the workers, who were mostly casual laborers, took a free-and-easy attitude toward the requirements. Workers wearing white or red decontamination suits were to be seen sitting and chatting in bars around the Seveso area. I was told by a medical man in the area that one day about ten workers wearing not decontamination suits and masks but T-shirts and shorts were seen playing soccer and kicking up dust in a part of Zone A where dioxin levels had been calculated at as high as one milligram per square metre. According to other people, the authorities attempted to deal with such flagrant safety violations by firing a few of the worst offenders, but instead of bringing about an improvement in work discipline the dismissals triggered a series of strikes by other workers, which were ended only when the offenders were rehired. The work of bulldozers piling up contaminated soil raised a tremendous amount of contaminated dust, which was often borne by the winds to points well away from the scene of the decontamination operations. Contrary to regulations, trucks passing out of the fenced-off area of Zone

A were not always washed down. Trucks taking loads of contaminated soil or foliage to the central part of Zone A often had no covering over the loads, and from these trucks clouds of dioxin-contaminated particles flew around.

When bulldozing operations in Subzones A-6 and A-7 had progressed to a certain point, the authorities also began bulldozing parts of Zone B where unacceptable levels of dioxin had been discovered. Because of the relaxed work procedures employed, these well-intentioned bulldozing operations apparently meant the raising of more dioxin-contaminated dust. Then, the removal by truck of this contaminated soil—and also of loads of branches and foliage from trees cut down in parts of Zone B—to the fenced-off central part of Zone A caused, according to critics of the program, the spilling of even more contaminated matter from uncovered trucks on the roads and streets en route. And all this activity meant the appearance of white- or red-suited figures at more of the bars around the expanded work area.

A similarly casual attitude toward safety measures seems to have existed during the attempts to decontaminate buildings within the fenced-off area of Zone A and in Subzones A-6 and A-7. A worker named Giorgio Bottiani, who for several months was part of a crew assigned to decontamination operations in buildings, said in a deposition made in October of 1977 that the precautions taken to protect the health of the cleanup workers were minimal, that the overalls issued to them ripped with ease, and that the masks they wore gave them insufficient protection from dust. Bottiani also stated that during the first two months of his employment the toxic waste washed from the interior walls of contaminated houses was simply poured into washbasins, bathtubs, and toilets. From there, it might be conveyed in part into the surrounding ground from drainage fields or it might be fed by drain into the Seveso River, down which it could course through other municipalities toward Milan and the Po. After a number of protests against these practices, the contaminated waste water was collected in concrete tanks placed outside the buildings being cleaned. Then either the tanks were taken by truck to be emptied elsewhere or the waste in them was pumped through hoses from the tanks into a field lying just within Subzone A-5.

Further serious problems arose out of the attempts to reduce the contamination of the area affected by the toxic cloud. One of the grimmest concerned the disposal of the carcasses of the eighty-one thousand domestic animals that by June of 1977 either had died of dioxin poisoning or had been slaughtered as a sanitary measure in Zones A and B and the Zone of Respect. Most of these contaminated carcasses were still stored in plastic bags or fifty-five gallon drums in the central part of Zone A. According to critics of the decontamination program to whom I talked while I was in the Seveso area, however, some of the carcasses were not taken to this already contaminated area for storage but were taken to Milan to be burned in the incinerators of slaughterhouses—incinerators that I was assured were equipped to burn animal waste only at comparatively low temperatures. Yet a temperature of at least eight to twelve hundred degrees Celsius, depending on the duration of the incinerating process, is required to destroy dioxin. Moreover, any low-temperature incineration of material contaminated with dioxin and dioxin precursors involves the risk of actually *creating,* by heating dioxin precursors, more dioxin than already exists in the contaminated material and releasing it into the air. And, of course, the transport of the dead animals in itself meant some risk of spreading dioxin contamination along the roads.

Resentment among people in the area against what they

saw as increased health hazards actually created by the authorities who were supposedly protecting their health was becoming marked. People in Zone B were complaining openly about the transport of contaminated material between one zone and another. Not only were trucks constantly carrying loads of contaminated earth and vegetation from parts of Zone B—in particular, Cesano Maderno—into Seveso for piling up in Zone A but trucks were also carrying vegetation cut in one section of Zone B to dump sites in other parts of Zone B. The councils of various municipalities in the area, including Seveso, fell to squabbling among themselves over the issue. The people of one municipality wanted no part of any other municipality's contaminated material. By summer, there was an air of revolt brewing in Cesano Maderno against the constant movement and spillage of contaminated earth and vegetation. A number of women in the community, indignant at the sight of so many trucks loaded with dioxin-contaminated material bouncing along roads in the area where their children played, banded together and placed themselves in front of some decontamination trucks, effectively stopping the traffic. In Seveso itself, more than a hundred women blockaded other routes used by trucks bearing contaminated material. This particular blockade lasted several weeks. The result of the demonstrations was to bring the trucking of contaminated materials to a virtual standstill through the whole affected area.

In Cesano Maderno, dissatisfaction reportedly reached the near-riot stage when, in June, a local medical officer, having ordered three cows belonging to a small farmer within the town limits to be slaughtered for laboratory analysis of the possible dioxin content of their internal organs, had the carcasses buried in a pit dug in an inhabited area within the municipality. Residents of the area where the cows were buried were so incensed that some of the men dug up parts of the carcasses and deposited the decomposing remains on the doorstep of the Cesano Maderno town hall. It was even said that a cow's head and entrails had been placed in the mayor's parlor as a further sign of displeasure with the way the decontamination program was being carried out.

Evidence of dioxin contamination throughout and beyond the area that was originally considered to have been affected continued to increase, even in places that previous measurements had indicated were dioxin-free. Late in June, traces of dioxin were found at a nursery school in Seregno, several miles to the east of Seveso, and that same month the discovery of dioxin in two schools in Nova Milanese, about five miles south of Seveso and eight miles north of Milan, forced the closing of these schools so that attempts could be made to decontaminate them.

In spite of such evidence, it appears that during the summer of 1977 the regional authorities did not carry out any systematic measurements of dioxin levels in schools in Zone B. Decontamination operations in the schools appear to have been carried out only sporadically and on the most impromptu basis. As the beginning of the new school year approached, it became clear that a considerable number of schools were in fact contaminated with significant traces of dioxin—even schools that earlier in the year had been examined and pronounced dioxin-free. "During the summer-vacation period, the responsible people in the region simply didn't seem to think it essential to clean up the schools before their opening," one of the physicians monitoring the health of the inhabitants told me. When the time came for the schools to open, hurried tests of dioxin levels indicated that more than a hundred schools in municipalities around Seveso were contaminated—only at low levels, the authorities insisted, but contaminated nonetheless. As a result, a hundred and twenty-four schools of a hundred and thirty-three tested had to remain closed until cleanup procedures and further tests could be carried out.

In October, it was discovered that a new prefabricated nursery school that the Swiss government had installed to the northwest of the ICMESA plant was contaminated with dioxin, and at levels as high as twenty micrograms per square metre. According to information given me by Dr. Gabriella Grimaldi, who has been working at Desio Hospital as a consulting psychologist dealing with evacuated Seveso families, earth used as fill around the school had been brought in by contaminated trucks and bulldozers, and she said that there were charges that some of that earth had originated in a contaminated part of Seveso. An even more disturbing allegation concerning loose decontamination procedures was reported to me by Dr. Grimaldi and confirmed by other people in the area. According to Dr. Grimaldi, after the ICMESA explosion the bodies of many of the poisoned animals had been stored, pending postmortem examination, in plastic bags in a gymnasium in the basement of the De Gasperi Primary School in Seveso. During this initial period, the school was closed. Then, after the carcasses were removed and some decontamination operations were carried out, the school had been declared dioxin-free, and it had been reopened for classes. Dr. Grimaldi told me that nineteen children at the school had subsequently developed chloracne, though medical examinations of the children which were made well before they attended the reopened school had shown no signs of chloracne among them. Indeed, according to Dr. Grimaldi, at the time of the ICMESA explosion most of the children had been away from Seveso on holiday and thus out of contact with the toxic cloud.

Even after the cleaning operations were undertaken and the schools were reopened, in the fall of 1977, there appeared to be no guarantee that with the constant traffic of pupils the schools would not become contaminated again before long. This prospect stirred up tremendous turmoil among the teachers at the schools, leading to a series of protest demonstrations and actions by the teachers' union, including teachers' strikes. Individual teachers demanded the right to be transferred at will from schools in various communities south of Seveso without penalty to their careers. When the Milan provincial education authorities eventually gave in and allowed such transfers, many of the students began insisting that if the teachers didn't have to remain on premises that were likely to be contaminated with dioxin the students should also have protection from exposure. They began strikes and demonstrations of their own, particularly in Cesano Maderno. Some of the older teen-age pupils commandeered a large high school, and one day in November several hundred students held a public rally in the central railroad station. The essence of their protest was not that they should be excused from attending school but that the health authorities should be compelled to take stringent precautions to protect the well-being of those who did attend school. What the students were demanding for themselves and their fellow-pupils was a series of regular health checks at school, including regular monitoring for signs of dioxin poisoning—checks that they declared the student body had never been given.

DIOXIN contamination was clearly continuing to show up outside the formally restricted area, and in such a way as to make almost farcical the official maps supposedly delineating its borders. To add to the confusion, there appeared to be no way of knowing whether newly discovered pockets of dioxin contamination in areas farther away from the site of the ICMESA explosion were there as a result of the mechanical transport of dioxin from the contaminated area or had been there since the explosion and had simply not been properly measured. One difficulty was that the original measurements of dioxin levels in the soil of various areas were made using samples taken many metres apart, and the levels shown in these samples were used to draw a picture of a hypothetical continuum of contamination of the terrain. But the assumption of such a continuum proved to be wrong: the dioxin fallout apparently descended in irregular fashion and, in many instances, varied not merely from one sampled spot to the next, perhaps fifty metres away, but almost from one square metre to another. As a consequence, inhabitants moving through an area that supposedly contained uniform and acceptably low levels of dioxin contamination might actually be picking up and carrying along with them much higher levels, from small pockets of highly contaminated soil or other material.

Another difficulty was that the process of analyzing soil samples was a complex one, requiring difficult chemical separation and extraction procedures and the interpretation of signals given by gas-chromatograph and high-resolution mass-spectrometry equipment; in fact, it demanded technical skills of a sort possessed by relatively few specialists in a very few laboratories scattered through the Western world. Most of the soil-analysis work arising out of the Seveso disaster had been assigned to people in the School of Pharmacy of the University of Milan; while they had the necessary equipment, the sheer amount of work involved and the pressure under which it had to be done must have strained their resources and made at least some analytical errors almost inevitable.

This was at least an understandable scientific difficulty. But what did most to increase the problems of coping with the aftermath of the Seveso disaster and looking after the welfare of the people in the area was the chaotic political situation prevailing in Lombardy, as in the rest of Italy. A bureaucratic morass seemed designed to engulf any urgent social endeavor. This was exemplified in what happened to the plans to handle the health problems of people in the area. Originally, Vittorio Rivolta, the regional health official, who is a Christian Democrat, was supposed to have complete control over the committees set up to supervise various aspects of the project: epidemiological studies and clinical work, the decontamination program, and the analytical work on dioxin levels in human and animal tissues. Political-party influence and standing played a great part in the composition of these committees, and it was not long before all sorts of disagreements and conflicts, some of them scientific but many of them political, divided not only the various committees but the various members of each. There were also financial problems caused by the tortuous workings of the ramshackle Italian bureaucracy. For example, a grant was made by the Italian National Institute of Health, in Rome, to the Lombardy region to provide the regional authorities with qualified chemists from the institute to carry out the necessary analytical work on dioxin contamination. But in practice it proved almost impossible to recruit chemists from Rome on any long-term basis, because most of the chemists were already obliged to have second jobs in Rome in order to subsist on their low salaries. It appears that those chemists who did come to Milan soon returned to Rome, because Italian regulations forbade their being paid a living allowance of more than some fifteen dollars a day even in as expensive a city as Milan.

Before long, the whole program for the protection of the Seveso people was seething with quarrels and machinations, not only in the ruling Christian Democratic Party organization but also in the Socialist, Radical, and Communist Parties. Almost every detail of the work of the committees became a party matter, with the minority parties being determined not to allow Rivolta and the Christian Democrats to exert permanent control over the committees. In May of 1977, therefore, Rivolta abolished the existing committees and put all the Seveso epidemiological, decontamination, and analytical programs, along with certain others, in the hands of a newly created Special Office, with headquarters in Seveso, at a seminary near the ICMESA factory. The purported aim of this reorganization was to make all the operations more efficient by centralizing them under one authority. The director of the combined operations, Antonio Spallino, who is also the mayor of Como, would report directly to the president of the Lombardy region, Cesare Golfari, and so would presumably keep control of the programs and the people working on them in the hands of the Christian Democrats.

The regional authorities said that the Special Office, which began operations in summer, would breathe new life into the attempts to deal with the Seveso emergency, but the immediate effect was to deprive the work of what continuity had been provided by a number of dedicated professional people and technicians. For one notable example, toxicological and other analytical studies that were being carried out by specialists at the Mario Negri Institute for Pharmacological Research, in Milan, were simply taken out of the institute's jurisdiction, and much of what these specialists had been doing was never handed over to any of the groups under the authority of the Special Office. Further, most of the available literature on dioxin contamination and toxicological studies in the field is in English, and only a small fraction of it has been translated into Italian. This presented no problem for the specialists at, for example, the Mario Negri Institute, many of whom had studied in England or the United States and were fluent in English. But now, because Spallino and apparently everybody else connected with the Special Office—not to mention the people in the regional health headquarters in Milan—understood not a word of English, the greater part of the technical literature on dioxin was simply inaccessible to the people directing the dioxin-decontamination operations. Spallino himself was a lawyer, and, so far as I was able to determine, had no particular experience in chemistry or medicine or toxicology. The disbanded committee on epidemiological studies was not replaced by another committee as such. Although one of the supposed virtues of the new Special Office was that its work would be carried out in Seveso itself, specialists could not be persuaded to leave their jobs in Milan and come to work in Seveso. In some instances, the hazards of working in Seveso may have been responsible for this attitude, but many of the specialists were dependent on political connections for their jobs in Milan, and they were unwilling to exchange their existing, often politically precarious, working arrangements for work that was potentially even more politically precarious.

The work that the new Special Office seems to have pushed forward with the most vigor was the bulldozing of contaminated soil in Subzones A-6 and A-7 and its removal to dumps in the fenced-off central part of Zone A, together with the cleaning and repainting of houses in the two lower areas in anticipation of the return of as many evacuees as possible by late fall. But these efforts seem to have been made at the expense of such necessary work as carrying out systematic health monitoring of people living in the entire affected or potentially affected area, and containing the spread of dioxin by human and motor traffic. From the time the Special Office was established, complaints were heard all over the area that funds theoretically made available for Seveso by the Italian central government and the Lombardy regional government weren't there when needed. As a result, much vital work supposedly being done under the Special Office either slowed to a halt or was never started, even when safety considerations demanded prompt action—work such as regular and thorough surveys and chemical analyses of dioxin levels at schools in the area and the necessary cleanup of school premises. Partly because of such difficulties, and partly because political bickering came to afflict every part of the handling of the Seveso affair, the Special Office staff and associated groups were plagued by the resignations of medical and other specialists—people who in trying to carry out their work had come to feel so frustrated by bureaucratic and other snarls that they simply gave up and left. With the disbanding of the epidemiological-study committee, most of the responsibility for monitoring the health of the people fell to a small and overworked group of physicians at the Desio Hospital.

The ever-mounting official confusion and disorganization had a marked effect on the attitude of the people of Seveso and the surrounding area, many of whom became skeptical of the usefulness of the programs meant to cope with the aftereffects of the ICMESA explosion. Increasingly, evacuees from Zone A turned their backs on the authorities, on the politics of the disaster, on the official advice being tendered, and, it seems, almost on the idea of dioxin itself. They had become so wearied by their difficulties that they wanted only to get back into their houses as quickly as possible, no matter who might say what about the hazards involved. Many of the people stopped cooperating with the health authorities, and simply refused to turn up for scheduled health checks.

For example, the original epidemiological program provided that people who had been or were about to be examined by clinicians would be given special health cards, which they could present on future visits to any clinics they might go to in the region, so that doctors treating them would be alert for symptoms related to dioxin exposure. "Most of the people simply threw away their cards," a physician who had formerly been associated with the health program told me. "Even many of the women who were pregnant at the time of the explosion just tore their cards up before they went to the hospital to have their babies. They were ashamed to admit that they had been exposed to dioxin on the day of the explosion." Moreover, many people shunned the official clinical examinations and relied for medical advice on their private local doctors, with whom, as it happened, the formally constituted health authorities had no effective communication about prevailing health conditions. In general, the Seveso people came to believe that, given all the conflicting official information offered about the dangers of dioxin, their own judgment about the hazards they faced was probably as good as any, and perhaps better. And this judgment often led them to resist the very idea of an imminent dioxin danger.

One person who has had regular contact both with the Seveso evacuees and with people in Zone B since the toxic cloud descended is Dr. Laura Conti, a Communist member of the Lombardy regional council, who is also the current head of the regional health and ecology commission, and when I was in Milan I had a talk with her. Her political position on the programs for handling the aftermath of the

Seveso disaster appeared to be a curiously mixed one. Though she displayed a sort of Party contempt for local officials who had not helped correct industrial conditions leading to the ICMESA explosion ("They shot Ghetti because he was stupid," she told me), she also, as a functionary of the Communist Party now acting in concert with the Lombardy Christian Democrats, seemed to have generally supported the actions of the regional authorities after the explosion. (She was one of the officials who voted to gerrymander the superstrada out of the prohibited zone in Seveso.) While she was deeply sympathetic to the plight of the Seveso evacuees and those living in the affected area, she also disdained the reaction that some Seveso people had to the health issue. When I made some mention of "the terrible experience" I thought the Seveso people must have been exposed to, Dr. Conti said, rather sharply, "It hasn't been such a terrible experience for most of them, because they never really understood the danger." She went on to talk about the attitude of many of the Seveso residents toward the outbreak of chloracne that developed in the area, mostly among the children. "You have to realize that in Italian folklore skin diseases are not considered as real diseases," she told me. "They have a saying in the Veneto area, where a lot of the Seveso workers come from—'An outbreak on your skin means that all's well inside.' " While the prevailing view of the effects of the explosion was a fatalistic one, she went on, the people in Seveso who came from the South, and might perhaps have been expected to take a more superstitious view of the disaster, in fact reacted in a more rational fashion than many of the northern people. "People from the South tended to see that danger really was there, and when they could they sent their children away after the explosion to stay with relatives in the villages they came from," she said. "The natives of Seveso and the people from other parts of the North were more of a problem. They tended to take the attitude that if children came into the world malformed, through whatever circumstances, it was God's will." But she felt that under the fatalism lay feelings of anguish and guilt. "Actually, some of the people who must now be considered at the greatest risk, because they lived closest to the ICMESA factory, feel the guiltiest," Dr. Conti said. "Just because they lived near the factory, they knew long before the explosion that chickens, rabbits, and other domestic animals were dying off right in their own back yards. And they would take the carcasses to the factory management and get ten thousand lire"—about twelve dollars—"for a dead chicken, provided that they promised not to report the death to Ghetti. Since the ICMESA explosion, these people have naturally found it difficult to face the thought that what they really sold was not a chicken here and there but the health of their own children. They can't bear it. And so such people try to insist that what they hear about dioxin can't be true—that dioxin can't be so harmful."

Dr. Conti told me that at a meeting held early in 1977 between Seveso people and representatives of the official health-monitoring groups to discuss the rising rate of chloracne among children of the area some of the Seveso people present maintained that it could not be true that dioxin was harmful, as was continually stated. According to Dr. Conti, they believed a story circulated by one of their number that during the war in South Vietnam the Agent Orange laid down by the American military in defoliation operations "made the bananas grow a foot long, so the rats grew a foot long, too." And some Seveso people at the meeting who were living in Zone B not far from the Zone A demarcation line claimed that since the toxic cloud descended rats a foot long had appeared in Seveso. "I asked them, 'Have you yourselves seen these big rats?' "

Dr. Conti said. "They said they had. They said they stayed up at night with sticks to ward off the rats from their homes. My own speculation about what may actually have happened is that because the wheat and oats that were growing on the small farms in Zone A at the time of the explosion had gone unharvested, the rats in the area had grown fat on them."

As time went on, some people's resistance to acknowledging the reality of the dangers of dioxin contamination seemed to increase, while others apparently felt increased indignation at the ineffectuality of measures supposedly being taken to minimize the dioxin threat. But there could be no doubt about the desire of the Seveso evacuees to return to the homes they had been forced to abandon, particularly in Subzones A-6 and A-7, and so strong did this desire become that attempts by anyone to demonstrate the dangers of dioxin contamination in the area tended either to be largely ignored or to be construed not as contributions to the strengthening of public health but as some subtle plot against the welfare of the community. Thus, attempts made by a so-called People's Technical and Scientific Committee, a largely leftist group, to warn people outside Zone A about both the immediate and the long-term dangers of dioxin contamination over a broad area around Seveso appear to have failed almost completely. "The real problem of the committee has been to get the people themselves to recognize the nature of the problem," the committee secretary, Antonio Chiappini, who until shortly after the explosion had himself worked as a laborer at the ICMESA factory, told me. Nonetheless, he added, he still had hopes that more people would attend meetings of the committee "and hear the message about dioxin."

But—if what I was told by toxicologists formerly connected with the epidemiological-study programs in Seveso was true—in addition to the reluctance of many of the people in Seveso and the adjacent area to face unpleasant facts about the spread of dioxin around them, and its toxic potential, there were signs of outright sabotage of certain scientific programs that had been undertaken to measure the possible bioaccumulation of dioxin in the Seveso environment. According to the toxicologists, when scientists planted experimental plots of carrots and other vegetables in the more heavily contaminated parts of Zone A, to determine whether dioxin in the soil might be incorporated into the structure of growing vegetable matter, some of the plots were vandalized again and again, the carrots being ripped out of the ground and removed. Similarly, when some rabbits and guinea pigs were placed in wire enclosures in parts of Zone A in experiments designed to measure the rate of absorption and accumulation of dioxin in the internal organs of the animals at differing contamination levels in the soil, scores of the experimental animals were stolen or killed on the spot. One of the scientists connected with these sabotaged experiments commented, "A faction within the Seveso population hates to have further confirmation of the dangers of dioxin, and has evidently been ready to use any means to stop experiments that might provide such confirmation."

Such incidents reflecting the underlying tensions and conflicts within the population remind one of Ibsen's "An Enemy of the People." The drama's protagonist, the physician Dr. Stockmann, having discovered that the water of the health spa for which his town has become well known is in fact dangerously contaminated, attempts to circulate public warnings concerning the nature of this health hazard. But, instead of receiving, as he anticipates, the gratitude of his fellow-townspeople for performing a valuable public service, Dr. Stockmann finds himself running afoul of local economic interests and forces of laissez-faire, for whom his information is highly inopportune. And such is the reluctance, growing into resistance and outright hostility, encountered by his attempts to convey the unpleasant truth that eventually the unfortunate doctor becomes not a local hero but a community outcast.

WHEN I drove back out to the Seveso area from Milan along the Milan-Como superstrada on a damp day after my talk with Mayor Rocca, it was through a mist rendered pungent and oppressive by the industrial pollution that permeates the whole area north of Milan. The landscape I drove through was dreary—smoking industrial installations interspersed with cement-block apartment buildings and stuccoed houses, mostly of recent construction and of unengagingly uniform appearance. No traffic signs were necessary to remind me where to turn off for Seveso. The presence of the most heavily contaminated part of the area touched by the toxic cloud was proclaimed on the west side of the superstrada by yellow-and-black signs reading "Zona Inquinata" ("Contaminated Zone") and by a tall yellow corrugated-plastic fence that extended from near the Seveso exit northward along the highway as far as I could see in the mist. I did not turn off into Seveso then but continued north. At one point, I noticed two children playing just off the edge of the highway and within a few feet of the yellow fence.

After about a mile, as the superstrada and the yellow fence curved to the left ahead of me, I glimpsed, looming out of the murk and within the prohibited zone, a complex of squat buildings and girder-surrounded distillation towers that I recognized from photographs as the ICMESA factory. A little farther along, I came to an elevated stretch of the superstrada, and I could see the ICMESA buildings much more clearly than before. I could also see, to the east of this complex, the circular outlines of the foundations of two large concrete silos that had been under construction when the explosion occurred. These foundations, I knew, were now used for storing some of the thousands of animal carcasses that had been gathered in the northern part of Zone A. I could also identify, from my memory of press photographs of the plant, the building housing the trichlorophenol reactor that had gone out of control, and, protruding from its roof, what I took to be the vent from which the plume of poisonous vapor and particulate matter had burst upward, to disperse in the wind as the toxic cloud. It was an eerie experience just to consider what was within that particular building now. Nobody knows precisely how much dioxin is still in the reactor, but by 1978 the vessel was estimated to contain three or four tons of trichlorophenol mixture, now in solidified form, including perhaps three hundred grams (two-thirds of a pound) of dioxin—an awesome quantity, considering its almost incomparably high toxicity. At the time of my return to Seveso, the building containing the reactor was considered so unsafe that only a few well-protected technicians had entered it since a few days after the explosion seventeen months previously. How and when, if ever, this chemical time bomb can be defused, or removed to any conceivably safe storage place, presents a question to which nobody has yet devised a satisfactory answer.

Once I had passed the northern end of the yellow fence, I left the superstrada and turned south, and after a few more turns I found myself at a big iron gate marking the entrance to the ICMESA plant, right on the edge of the prohibited zone. The gate was guarded by several carbine-bearing Italian Army soldiers, who were standing about casually chatting. They were wearing disposable blue decontamination suits of some paperlike material, and also white face masks, but not on their faces. Two of them were wearing the masks on the back of an arm, nestled on the elbow. I continued south, along the perimeter of the upper part of the zone. I was unable to see anything more inside the tall yellow fence; later on, I was told that the abandoned houses in the upper part of the zone lie amid an expanse of vegetation that has grown so uniformly thick as to obliterate most of the existing features of the landscape, such as roads, pathways, and garden fences. Most of the furniture that remained in these houses after the earlier comings and goings of the inhabitants was apparently removed by decontamination crews during 1977 and thrown into pits that had been dug in Subzone A-5 and lined with concrete. I learned that the large piles of plastic bags containing animal carcasses that had been stored aboveground in A-5 were being buried in other lined pits in A-5.

I continued driving south to a point beyond the southwest corner of the zone enclosed by the yellow fence, and then, spotting a little café called the Trattoria Savina, I stopped off and ordered a cappuccino at the bar. At a few tables in the café, men were sitting quietly playing pinochle. While I was sipping my cappuccino, I got into conversation with a short, grandmotherly woman behind the bar, who, I learned, was a co-owner of the café. Her name was Agnese Galimberti. She told me that the street the Trattoria Savina was on was the Corso Isonzo, which lay just outside the Zone A boundary. She said that after the ICMESA explosion and the evacuation, the regional authorities had closed the Seveso entrance and exit from the superstrada that connected directly with the Corso Isonzo and kept them closed for about fifteen months, and that, since a good part of her business in the past had come from passing traffic, she had

had hardly any customers during all that time—and, of course, none from among the people who had been evacuated from Seveso. She said it was only now, with the reopening of the superstrada connection and the return of many residents to their homes in A-6 and A-7, that her business was returning to normal.

I asked her how seriously people in the area seemed to regard the hazards of dioxin.

Signora Galimberti replied, "Rather than think about dioxin, they just think about going back to their homes." She went on, "Getting ill—how do you know it's dioxin if you're ill? People died before dioxin came along, too! Nobody knows anything about what dioxin does or doesn't do. A few people may be afraid to go back, but only a few. The others are happy. Did you know that the Archbishop of Milan is coming here to Seveso on Saturday to bless all the houses the people have been moving back into? Oh, yes! This means he's giving his approval to the whole thing."

When I left the Trattoria Savina, I drove east along the Corso Isonzo into Subzone A-6 and among the houses that Seveso evacuees were reoccupying. The area gave the impression of a building project that had been just about but not altogether completed. Much of the ground was still unadorned by grass, trees, and bushes, and the side roads were newly oiled. Most of the houses were two-story detached dwellings of stuccoed cement block with red tile roofs. Only two or three people were about on the streets. Then I came to a turn that brought me within about eight feet of the yellow fence enclosing the zone. Right on this corner was a solidly built house, with an as yet unpainted stucco finish of a sort of buff-rose color, and with the usual red tile roof. Outside the house stood a small cement mixer, and near the cement mixer a man dressed in work clothes was building a low cement-block retaining wall. I stopped, got out of the car, and introduced myself. He told me that his name was Mafaldo Sartor, that he was forty-eight years old, and that he was a construction worker "on the artisan level" and a contractor. He showed me into his dining room, apologizing as he did so for the unfinished condition of the place. He and his family had started to move their belongings back only a week ago, he said, and a lot of work still had to be done. He said that to him his dwelling was "a beautiful house;" long before the ICMESA accident, he had built it with his own hands over a period of a year and a half, working Saturdays and Sundays. I could see that the carpentry around the place had been done with great care. The dining room was a good-sized one, with a large table, over which hung, as a sort of centerpiece to the room, two ornate chandeliers. Along one wall was a large walnut breakfront, heavily carved, with glass doors. We sat down at the table, on which stood a bowl of flowers, and we were joined by Signor Sartor's wife, Edda; their daughter Maria Grazia, who was twenty-two; and their son Claudio, who was eleven.

Signor Sartor said that the whole family was pleased about getting back home after a long dislocation—an experience he described as *"bruttissima."* Signora Sartor said that when they began returning their things to the house, "just from the emotion of moving back home I couldn't sleep all night—it's something you can't describe." Signor Sartor said that when the ICMESA explosion occurred the family surmised that something must have caught fire at the factory, because of the cloud and the smell that accompanied it. He said that they had remained in the area twenty-two days before being evacuated, and that for the first week they had eaten fruit from their trees and vegetables from their garden—until the authorities told them not to. When they eventually left their house, for a motel at Assago, forty-six kilometres away, that was a terrible experience, he said, "but living in the motel where we were evacuated was worse—just one room for the whole family, and it was like living in a psychiatric ward." He went on, "The health authorities wanted to kill our dog, Pipperi, but we hid him and managed to take him away with us. The dog had started to feel bad, and had been vomiting and had diarrhea."

"I'll never forget all the cats, and the way they died in the days after the cloud came down!" Signora Sartor said. "There were seven or eight cats right around our house. Their eyes swelled up and came right out of their heads. They got very thin, and as soon as they died their flesh seemed just to burn up. In three days, only the skeletons were left—it was as though the bones had been picked clean. And the birds—they just fell to the ground out of the air. I had some two-year-old hens. The day after the accident, they stopped moving and laid no more eggs. They were finally taken away from us by the health authorities. A neighbor of ours had chicks that had hatched. They all were deformed and died."

As for the health of the family itself, Signor Sartor told me that, for whatever reason, he lost twenty-six pounds in two months after the explosion. Fortunately, the children did not develop chloracne. But both he and his wife became terribly nervous and suffered greatly from worry—"a terrible time of stress that none of us will ever forget." He said that one of the things that had distressed the family most was the invasion of press, radio, and television reporters into their home and the homes of other residents, and the relentless way these reporters had pried into the lives of those who were being evacuated. "All the media people came here from England and France and Germany. And some of them were saying out loud among themselves—and I could understand it, because they happened to be speaking in Italian, to a journalist from Milan—'Just think, in two years all these people will be dead.' And they were standing there saying to one another, 'They'll just fall down like overripe pears.' That upset us the worst."

It seemed clear that in the months after the evacuation Signor Sartor had managed in one way or another to visit his house from time to time, even though doing so was officially forbidden, for he said that thieves had stolen things from his home on three different occasions, making off with clothing, bedding, rugs, and tools. He said that almost all the abandoned houses in Zone A had been looted. During the time they were away, his family had had only one desire, and that overwhelmed all other considerations: to get back into their house. He himself had very little sympathy with the various demonstrations that were held locally from time to time concerning the dangers of dioxin; he thought they were "politically oriented" and were probably organized by outsiders. The net effect of these protest demonstrations, he declared, was only "to slow down the people's return to their homes in Seveso by perhaps five months." He expressed no bitterness against the Givaudan or Hoffman-La Roche Corporations. He referred to the owners of the ICMESA factory only as "Switzerland." "Switzerland has paid us for the furniture that was taken away and buried as too contaminated," he said. "Switzerland has also paid for the cleaning of the houses. And they've given us compensation to cover most of our other losses, too."

About the supposed dangers of dioxin he said, "We don't really know what may have happened to the people's health. One authority will say one thing, and another will say something else. As for me, I believe what I see. I've heard that dioxin isn't as dangerous as it's been made out to be. We're back home again—that's what counts for us. Tonight will be the first time we have slept in our home in all these sixteen months. The dioxin story is finished for me. On Saturday, the Archbishop of Milan is coming here to bless all the houses. Including this one. The Archbishop himself will say bye-bye to dioxin."

THE next day, on my way out to Seveso again from Milan, I stopped off in Cesano Maderno to look around the center of town. It was the noon hour, and almost everything was closed. Here and there, on side streets, one could glimpse what Cesano Maderno must have looked like even as recently as a quarter of a century ago—courtyards surrounded by old buildings, many of which must have been stables or sheds for livestock or feed. But it seemed that nearly all such buildings had been torn down and replaced by apartment buildings or stores. The place had a prosperous air. I saw one or two furniture showrooms along the main street, with pieces that obviously weren't to be acquired cheap. Walking along one street, I spotted a pasta-maker's shop that was open. I went inside, introduced myself to the proprietor, and explained that I was just passing through. The proprietor, a man of about fifty, stood at a clanking machine that resembled a printing press, turning out sheets of meat ravioli. He told me that he made meat ravioli in the morning and spinach ravioli in the afternoon. I said I was on my way to Seveso to talk to people there about their circumstances and the question of dioxin. The last word was hardly out of my mouth before the pasta-maker waved both arms above his head as though he were fighting off flying objects. *"Diossina! Diossina! Diossina!"* he cried out, in an aggrieved voice. He said that the dioxin issue had brought him nothing but misery. At the time of the ICMESA accident, he said, he had been operating a pasta shop in Seveso, which was then his main place of business, as well as the one in Cesano Maderno, and that a couple of weeks after the descent of the toxic cloud the police had come to his Seveso shop and told him to stop making pasta, because the flour might be contaminated with dioxin. He told me that he was still allowed to make pasta in his Cesano Maderno shop, but that "if the customers don't see you producing it fresh, you get only half the price." He said that he had been refused compensation for the loss of business caused by the toxic cloud, and that his business had suffered so much that he had had to lay off four men, who badly needed the work. But some families who were moving back into Subzones A-6 and A-7 not only had their houses completely renovated free but had got ten or twelve million lire in compensation as well—"a good two years' salary." He went on, "There's so much corruption over this Seveso affair. People are getting rich on it. There's been a lot of funny business going on. What if Switzerland comes to the authorities and asks, 'What have you *done* with all the money we've been required to pay out for resurfacing roads, and so on?' How could they answer?" He said that all the bigger woodworking factories in the area saw to it that they themselves were well taken care of. "Lots of carpenters, craftsmen, the little fellows—they got nothing from Switzerland. But the big fellows got plenty. One got three hundred million lire in compensation. That's the way things were."

IN Seveso again, I dropped by the Special Office, on the western edge of Zone A, near the ICMESA factory. I was hoping to get an appointment with the head of the office, Signor Spallino, to discuss the current situation in Seveso. I had been trying for many days without success to make contact with Spallino. I had managed to talk several times with his principal deputy, and he kept telling me that Spallino was busy but that perhaps a meeting could be arranged.

At one point, Spallino's deputy had told me that I would probably be able to catch Spallino the next day in Milan, where he was going to "see the minister." The deputy advised me to stay by the telephone in my hotel room that morning; he would arrange an interview, he said, and would call to give me the details. I had stayed in my room all morning waiting for his call, but it never came. I was unable to reach the deputy by telephone. At noon, further action was made impossible by the arrival of the Italian lunch hour, which lasts until three o'clock—and after that, of course, it was too late to see Spallino. That's how things tend to go for anyone dealing with officialdom in Lombardy. I had also been trying for quite a while to get in touch with the press officer for the Special Office, a man named Sergio Angeletti, and when I finally did reach him I hastened to make an appointment to see him for a few minutes, mostly in the hope of arranging a meeting with Spallino.

I found Angeletti in a small room on the second floor of the Special Office. He was a stout man, with a long, bushy beard and with fringes of long hair at the sides and back of a bald pate. His manner was harried. He told me that Spallino was still very busy, but promised to try to get me an appointment with him and to let me know as soon as he succeeded. We talked for a few minutes. In answer to my questions, Angeletti said that soil-decontamination work in the entire area affected by the toxic cloud had ceased months earlier. "The people don't want to hear any more about decontamination work," he said. "They don't want dioxin removed from their soil. In one municipality, they refuse to accept the soil bulldozed from another municipality. The people don't want to see trucks and bulldozers—they think that if dioxin is in the soil, that is the best place for it to stay." He gave me a few figures—how many families were moving back into A-6 and A-7, and so on. He said that he was a journalist by profession, and that he had come to work for the Special Office thinking that he could contribute a public service. But things were difficult, he said. "I felt in coming here that it would be a good thing if a job like this could be filled by someone who wasn't a politician," he told me. "I wanted to play no political games. But, you know, they're nearly all Christian Democrats on top in this region, and I've had the impression that my lack of political motivation is proving to be a worry to the people around me. Every day, I realize more and more how politically motivated everyone at the Special Office is." Reaching for a sheet of paper, Angeletti drew for me some organizational charts with little boxes. He told me things were getting complicated for him personally. There was intrigue everywhere. He pointed to the boxes on a chart he had drawn and said that one faction was trying to have him shifted, in what was theoretically a promotion, to the position of "coördinator" in the Special Office, but that that was really a move to lessen his contact with the press. Everything was politics, he said again. I told him I'd keep in touch with him about the talk with Spallino.

THAT evening in Seveso, I attended an informal meeting of a small group of businessmen who had been particularly active in opposing the construction of a giant incinerator in Zone A. The meeting was held in the office of Aurelio Lunghi, a tax accountant, who was a man of some stature in the local business community. I had been invited to attend by Luigi Campi, a well-to-do importer of timber, whose business takes him to such faraway places as Gabon and Australia in search of woods to be used as veneer for the Seveso-made furniture.

At the meeting, Lunghi offered the opinion that the seriousness of the chloracne problem in Seveso and the surrounding area had been "grossly exaggerated" in the press, and said that this probably applied to the dioxin question in general.

Campi agreed. He said that the problem was "not as large as people have made out." Its seriousness had been exaggerated for political and monetary reasons, he went on, adding, "We are, after all, in the country of the *bustarelle*"—the "little envelopes," a reference to the prevailing system for bribes for getting things done. "And the bigger the problem is made to seem, the bigger the amount of money to solve it becomes, and the bigger the number of sharks who are there to get their share."

Nobody present seemed inclined to dispute Campi's statements.

The consensus appeared to be that the construction of a high-temperature incinerator in Zone A to destroy dioxin would, besides posing a further danger to health in the community, compound the economic difficulties that had already been suffered by businesses in the area. There were denunciations of the "ineptness and arrogance" of the regional authorities concerning the decontamination program. One after another of the businessmen spoke up to point out particular disadvantages of an incinerator.

"Just think. They talk of burning three hundred thousand tons of soil—and for only a few grams of dioxin."

"It sounds like a mad thing. To do all that, the incinerator would have to be the size of Milan Cathedral!"

"It would take them fifteen years to incinerate all the material."

"Consider the enormous running costs."

"We know that since the ICMESA explosion the Lombardy administration would like to see a big incinerator built for taking care of *all* industrial wastes in the whole area—not just dioxin. So they pick on Seveso, saying to us, 'You need to get rid of your dioxin anyway.' "

"The authorities think they can get the incinerator for

nothing—they would make Hoffmann-La Roche pay for it."

"And bring all their wastes to Seveso to burn—not just for fifteen years of burning but for an indefinite period after that. And ruin Seveso forever."

"The risks are absurd."

"I was born in Seveso, and I want to die in Seveso. But not from dioxin. I am a white-collar worker. There was a time when I was a partisan, and we partisans knew how to act when we had to. Recently, I told a friend, 'If they put an incinerator in Seveso, we'll use dynamite! We'll use bombs!' "

Some time after the meeting got under way, Mayor Rocca showed up. He smiled at everyone as he came in, but I had the feeling that he hadn't been invited. He had once been in favor of the incinerator, but he now left no doubt that he had changed his mind. After he had aired his views on the subject, he asked the assembled businessmen why the group had been organized.

"Because you didn't do your job," a man sitting next to the Mayor told him tartly.

Mayor Rocca then reiterated that because there had been no lasting health problems as a result of the explosion the incinerator was unnecessary. The talk turned to the question of the spread of dioxin from the most polluted zones to other communities. On this point, the Mayor turned out to have a view of his own. "The wider dioxin spreads around, the less of it people are exposed to," he said.

Hearing this, I reflected that although the people in the ever-expanding areas affected by the apparently continuous transport of dioxin through the whole region might not appreciate what was drifting their way, the Mayor's view wasn't so very different from the position taken by the Dow Chemical Company in the United States: that although dioxin might be one of the most toxic substances known to man, if you spread it around enough—for example, in 2,4,5-T spray operations every year over millions of acres—any hazard to human beings in the total area covered may be reckoned as being shared by everyone there without discrimination, in a sort of democracy of risk.

THE next day, I decided that if I was ever to manage to see Spallino I would have to dispense with niceties. Just before the noon hour, I drove to the headquarters of the Special Office, more or less barged in on Angeletti, and told him that I had to speak with Spallino. I found Angeletti even more harried-looking than before. He said that the Special Office was in a state of crisis. That very morning, about nine-tenths of the staff had sent a letter to Spallino offering their resignations. According to Angeletti, these people said in the letter that their working conditions had become impossible. Decontamination workers had been bringing an increasing number of lawsuits against the Special Office and the regional health authorities, claiming injury to their health as a result of dioxin contamination, and in these suits they were naming, as co-defendants, personnel of the Special Office, claiming that the people there were negligent in protecting the workers' health. Angeletti said that the staff of the Special Office found themselves defenseless against these accusations. "They have no protection on the part of the politicians" in the regional or provincial government, he said, nor did they have any protection against legal attacks against them instigated by the workers' unions. "Now every worker who has taken part in the decontamination program and has developed a red spot goes to the lawyers," he said, "and makes legal complaints citing the Special Office people as culprits." The staff people were demanding protection against such negligence suits, he concluded, and the place was in an uproar.

Rather hardheartedly, considering the circumstances, I said that I still needed to talk with Spallino. Signor Angeletti sighed and said he would do what he could. He left his office, and returned after a few minutes. He said that Spallino had to go to Milan but that he intended to have a quick lunch first, and he could hear my questions while he was having lunch. I said that that would be fine. Angeletti and I went out to a nearby trattoria to get together with Spallino. He proved to be a wiry-looking man in his forties, who was a sharp dresser. At the trattoria, which had the inevitable lunchtime complement of card players, Spallino had picked a table from which he could see a television set mounted on one wall, which was showing an Alpine slalom race. Spallino must have been a sports fan, because he kept an eye on the television screen during a good part of the interview.

I began by speaking about the stress that the Seveso people had been subjected to. Spallino said that the people who were moving back into their homes were naturally glad to be able to do so but that "the unknown scientific aspects" of dioxin had indeed caused a great deal of anxiety. There had been all sorts of demonstrations, he said, and under such circumstances there was always a component that was irrational. Although there might still be considerable "disorientation and confusion" among the public concerning dioxin, the anxiety aroused was a natural consequence of a traumatic situation, and, given the situation, the returning evacuees were reacting well.

I attempted to concentrate my questions on the relative risks for the evacuees returning to Subzones A-6 and A-7 and those living in Zone B. After all, I pointed out, the people of Zone A had been evacuated well away from the zone for some seventeen months, and thus had presumably been relatively free of continuing exposure to dioxin, whereas the people in Zone B had been exposed to dioxin contamination, even if at relatively low levels, during that whole period; in view of the extreme toxicity of dioxin, might it not have been a wiser course for the authorities to evacuate a much larger area, including Zone B, for the protection of the long-term health of the inhabitants?

Spallino replied that "all aspects" had been taken into consideration in the evacuation of the population. "The social texture of the whole area would have been much more powerfully affected than it was if Zone B had been evacuated, too," he said. "When the authorities decided not to move the people away from Zone B, it was thought that it was better that way; if they had been moved elsewhere, they might not have been contaminated with dioxin but they might have become contaminated with something else." He conceded that the failure to evacuate Zone B, and the constant traffic in and out of it, had no doubt increased the mechanical transport of dioxin around the general area, but he didn't believe that that was so important. As for the findings of dioxin contamination in schools in Cesano Maderno and elsewhere in the area at the beginning of the school year, he said that the instruments used in analyzing particulate matter taken from the schools appeared to have "gone mad." He found such findings "anomalous," he said, considering the fact that "measurements taken months previously in many of the same schools showed no evidence of dioxin." Without actually saying so, Spallino indicated either that extraordinary failures in measurement procedures had been involved or that dioxin had been introduced into the schools by means other than unintentional transport. He seemed to stop just short of saying that the high dioxin-level readings in materials from the schools could have been the result of deliberate sabotage for political purposes. (Later on, when I talked with Professor Silvio Garattini, the director of the Mario Negri Institute, about the newer, positive readings for dioxin at the schools, he suggested that they could be due to improved analytical techniques. But the atmosphere of suspicion and distrust hanging over much of the scientific work was such, apparently, that even a man normally as unexcitable as Dr. Garattini could suggest to me that the possibility "must be carefully considered" that someone might have carried dioxin contamination into various schools around Seveso at the beginning of the 1977 school year "just to make confusion.")

I asked Spallino about the current status of the proposal for the building of an incinerator in Zone A. He indicated that, apart from purely technical questions, the broad opposition of the local inhabitants to the idea of the incinerator made it clear that at present the scheme must be considered, for practical purposes, a dead issue.

I went on to ask Spallino about the aftereffects of the ICMESA explosion on public health and about the epidemiological findings so far.

He said that concerning the epidemiological aspects of the situation I should talk to the doctors at Desio Hospital, who would be able to provide me with the information I needed. Then, looking at his watch, he said, "I must go to Milan."

WITH the help of a scientist friend in Milan, I arranged for an interview with a group of four doctors at Desio Hospital that same evening. All were young men, but they seemed to be tired and dispirited. I told them that Spallino had suggested that I apply to them for information concerning current epidemiological studies being made in and around Seveso.

Acting as spokesman for the group, one of the doctors

said that proper evaluation of the epidemiological situation was impossible. "This is a small staff here, trying its best to keep up with the clinical work to be done, and, frankly, it is not the purpose of this staff to examine the prevailing situation from an epidemiological point of view," he said. He explained that since the epidemiological committee had been disbanded and the Special Office established, the group at Desio Hospital had made repeated requests that the Special Office provide qualified epidemiologists, without result. "We have asked for epidemiologists to be sent from Rome, but nobody has been sent so far," the spokesman said. "The Special Office has never even received any acknowledgment of a letter of request sent to Rome." As for the ten medical people who had been provided by the regional authorities to help the Desio Hospital group with the clinical work to be done among the Seveso population, both people from Zone A and people living in Zone B, only five of the ten still remained with the group. Unfortunately, the spokesman said, the Special Office, instead of bringing order out of a chaotic situation, had grown a bureaucracy of its own, which had had the effect of compounding the disorder. "Of five large programs outlined by the Special Office to deal with the problems of Seveso, the authorities have appointed a chief for only one, and that has to do with the physical rehabilitation of contaminated buildings," he said.

I asked the doctors what impression they had of the effects of dioxin on the population.

Their spokesman said that in general, except for the chloracne outbreaks among young people, there appeared so far to be no great clear-cut effects on the health of the population as a whole. Even the figures that the group had managed to assemble on the incidence of spontaneous abortion within the community appeared inconclusive, he said, but he cautioned that it was most difficult to generalize from the statistics gathered, because the study was hampered by a lack of information on abortion rates in periods prior to the ICMESA explosion; besides, many spontaneous abortions had undoubtedly not been recorded as such, or perhaps even noticed by the mothers if they occurred early in the pregnancy. He said that in view of the widespread organic damage observed in animals and the number of animal deaths one had to assume the possibility that similar damage could be done in man. He said that in one part of the affected area forty cows were being bred at the time of the accident. They had been fed grass grown in fields near the ICMESA factory, and after the accident they continued to be fed grass originating from those fields. Of thirteen pregnancies among them, there were ten spontaneous abortions, and one of the aborted calves was malformed. Of the three calves that were carried to term, only one survived more than a short time after birth. Animal deaths had occurred as far south as Nova Milanese, he told me. He said that because of the severity of the observed effects of dioxin exposure in animals it was necessary to regard the risks to man as very serious, even though indications so far were that the actual exposure of most people in the area to dioxin as a result of the explosion must have been lighter than originally feared. "They have found dioxin in the livers of animals in the area that seemed to be healthy, so one must be cautious about potential problems among apparently healthy human beings," he said. The spokesman went on to say that adverse symptoms other than chloracne which were apparently connected with dioxin exposure were showing up among the population "here and there." For example, in one survey made of people who had been living within the upper half of Zone A and of people who had continued to live outside Zone A but on the same general east-west band, thirty-five per cent were found to have enlarged livers—a significant development, it seems, in view of the fact that liver damage is one of the prime effects of dioxin poisoning.

I said that I was interested in the opinions of the doctors concerning the possible long-term hazards to people in Zone B from continuous exposure to low levels of dioxin-contaminated material that had either been present since the accident or been transported by wind and rain and the movement of vehicles and people about the area.

The spokesman said that it was undeniable that dioxin was moving about in the area. He discussed the implications of a report by Dr. James Allen, of the University of Wisconsin Medical School, which the Desio doctors had read in Italian translation. As I have previously mentioned, Dr. Allen and some associates studied the effects of low levels of dioxin on nonhuman primates. In the study, eight rhesus monkeys that were fed a diet containing dioxin at a level of five hundred parts per trillion for a year became anemic and developed various toxic symptoms within six months, and then, between the seventh and the twelfth month, five of the eight died. And in another study by Dr. Allen, in which rats were fed a diet containing extremely low levels of dioxin, in the parts-per-trillion, and very low levels, in the parts-per-billion range, over a period of more than a year, almost forty per cent of the animals developed neoplastic changes—a high rate of damage that, according to Dr. Allen's report, "suggests the carcinogenic potential of the compound." The results of the Allen studies seem to suggest—contrary to the implications of the public position held for so long by Dow Chemical concerning the allegedly insignificant hazards involved in the continued widespread use of 2,4,5-T—that exposure to low levels of dioxin over a long period may turn out to be at least as harmful as short-term exposure to higher levels. Interestingly, a fairly recent study by Dow's own toxicologists on the long-term effects of dioxin introduced into the diet of rats tends to agree with Dr. Allen's findings of neoplastic changes in rats at continued low, parts-per-billion exposure, although the Dow study did not find comparable neoplastic changes occurring in the even lower parts-per-trillion exposure range. None of the Dow scientific studies appear to have changed Dow's long-standing corporate position on the safety of 2,4,5-T.

The doctors at Desio Hospital to whom I talked seemed most concerned about the implications of the Allen studies for the health of the people in Zone B. "To act in accordance with Allen's findings in this area would certainly mean a vast program of decontamination, if not evacuation, but such a program runs against the political and economic situation here," the spokesman for the doctors said. Only time would tell what the cumulative effects might be. He ended our talk by declaring that the problem of dioxin was a worldwide one, and that in view of its importance it was unfortunate that "the international scientific world hasn't been involved in the problems of Seveso as it should be." He added that those trying to cope with the medical and toxicological problems posed by the Seveso accident could use all the American scientific help they could get. Hearing this, I found it painful to say what came to my mind: that the American scientific community, which has been aware of the extreme hazards of dioxin for years, hasn't even been able to prevail on the United States government to put an end to the sale or use of it or the continuing exposure of its own population to it—that, indeed, the government itself is continuing to use the 2,4,5-T herbicide in its own widespread spray programs in the United States.

Later on, I tried to supplement what I had learned from the Desio Hospital group with information from other scientific sources in Milan. I was told that, besides chloracne and liver impairment, unfavorable signs appearing on a scattered basis among the population in the area in and around Seveso included diminished white-blood-cell counts; neurological disorders of various kinds, such as blurred vision and loss of conduction time in peripheral nerve tissue; reading tiredness among schoolchildren; disturbances of the endocrine system; and disturbances in the normal functioning of the enzymatic system. Between 1976 and 1977 the incidence of infectious diseases among the affected population tripled, and this striking increase has been interpreted by some specialists as an indication of the capacity of dioxin contamination adversely to affect the functioning of the immune systems of people living in the Seveso area.

And there were certain other unfavorable signs, which did not necessarily get reported in official medical accounts concerning the condition of people in the area. For example, I was told that one of the Seveso doctors to whom many local mothers took their children for examination, rather than to the official health-monitoring groups, was extremely concerned about a marked increase in episodes of convulsions among infants.

Since the ICMESA explosion, the birth rate in Seveso and the surrounding area is reported to have dropped sharply. Some physicians attribute this decline to the practice of birth control by the women as a precaution against bearing possibly defective children. However, in 1978, Dr. Allen and his associates completed a third study on the effects that dioxin and related compounds have on nonhuman primates. Of eight female rhesus monkeys fed TCDD in their diet at levels of only fifty parts per trillion for seven months and then mated, four of the females conceived but aborted their fetuses, and two did not conceive, although they were mated on repeated occasions. And while all eight animals used as controls in the study conceived and gave birth to normal infants, only two of the dioxin-exposed animals conceived and were able to carry their infants to term. Such data suggest that the reproductive difficulties associated with dioxin exposure in the Allen study and the drop in the birth rate among Seveso women may be part of an over-all pattern.

ON Saturday, I drove into Subzone A-6. As usual, the weather was raw, damp, and misty. I parked near the house where I had interviewed Mafaldo Sartor, then walked a block west beside the grim yellow fence. Knots of residents of the reoccupied subzone were standing around in a narrow road of gleaming black asphalt which ran north and south through rows of freshly painted red-roofed houses. Most of the people were dressed in their Sunday clothes. They were waiting for the Archbishop of Milan. I joined a group of residents halfway down the street. They were in a state of subdued excitement. One of them told me that, no matter what the weather, it was a beautiful day for all of them; now that they were back home, they could feel that their difficulties were mostly over. One man told me that, considering all the circumstances, "the people say they can't complain—they've been completely reimbursed by Switzerland for their losses." Still, he went on, there was no way of completely making up to the people for what they had suffered since the ICMESA accident. Another man in the group said to me, "Before the dioxin, this wasn't an asphalt street. Switzerland finished it." He said that at the time of the explosion some of the newly constructed houses were still in an unfinished state but that "Switzerland" not only had had them cleaned but in some cases had had the unfinished work completed.

After a few minutes, there was some commotion down the street, and soon I saw a small procession. First came two red-robed, white-surpliced altar boys swinging censers, and then the robed Archbishop of Milan, wearing his red hat, with two priests bringing up the rear. The procession turned onto the street where I was standing. The spectators crossed themselves. The Archbishop, with his aspergillum, made motions of sprinkling holy water here and there on the new asphalt provided by "Switzerland." A few moments later, still preceded by the altar boys, he turned in at the entryway of one of the houses, where the occupants waited reverently for the blessing to be given inside. One of the two priests, going over to a blue Fiat parked at the curb, opened the trunk and took from it a small armful of plastic boxes containing crosses, presumably for presentation to the people in the households being visited.

A while after Cardinal Colombo had left one of the houses he blessed, I called upon the people inside. The place had a festive look—lots of roses and other flowers carefully arranged in vases. The house, which was solidly built, belonged to Mario Scozzin and his wife. Scozzin himself wasn't there when I entered; he had already gone back to work. He was a building contractor. Signora Scozzin was sitting in the dining room with a few friends, having coffee. She said that her husband had built the house himself, with the help of a couple of workmen, and that she and her husband had done the finishing work together. "In order to build our home, we had denied ourselves so many things, and it was so important to us," she told me. "We had been living here only two years when the accident happened. One day we felt ourselves rich and the next day it was as though we were the poorest people on earth. We saw the birds falling out of the sky, the cats and chickens dying—we'll never forget. But now we're back to our home after all these long months of misery, and we are happy again at last." She said that, remembering what she had seen, and the chloracne that some of the children had developed, she could hardly believe that the dangers of dioxin were invented, but she and her husband were not afraid for themselves or their children now that they had moved back into their home, which had been cleaned and repainted. She was a little more careful around the house than she used to be, she said. For example, she would not let dust accumulate on furniture from one day to the next, as she used to, but made sure it was got rid of quickly. Yet now she felt that they could lead a normal life.

LEAVING the Scozzin house, I thought I would wind up my visit to Seveso by stopping off at a couple of points where I might sense what people living outside Zone A felt about the current situation. I drove up to a group of apartment houses about five hundred yards west of the Zone A boundary and within the Zone of Respect, where, I had heard, there had been complaints about the lack of decontamination measures by the authorities. In an apartment occupied by the Brambilla family, I met a group of women—Signora Romilde Brambilla, together with several of her neighbors. Signora Brambilla explained that she had been unwell and had been advised by the Seveso municipal authorities to go to the seaside to recuperate, that she had gone to the Ligurian coast, near Genoa, and that her husband had gone at his own expense to join her.

One of the neighbors, Signora Maria Veneziano, informed me that after the accident the people in the apartments had told the authorities that they wanted to be evacuated from the area along with the other people. "We're all afraid of staying here," she said. "We are right on the edge of Zone A, we are right near the ICMESA factory, and we feel that it is unsafe here. We've been to the Mayor about it, and we've been to the regional authorities, and they say the dioxin levels here are low enough so we can live without worry. It's impossible! Can we be so near to the most polluted part of Zone A and have no problem?"

Another of the neighbors, Signora Lucia Buquicchio, said that since the explosion she had developed liver trouble and had had to have a kidney removed.

Signora Brambilla said that a number of the people in the apartment house had developed blurred vision, which made it difficult for them to watch television. "So many people seem to have disturbances here on the block," she said. "It's not normal at all. They get dizzy spells and headaches, and there is still a lot of diarrhea. The doctors say the people's platelet counts are very low. They have skin problems, and when they go to the sea and it's windy the itching is terrible—*un tormento!*" She said she was convinced that there were a number of people on the block who really weren't well, adding, "But they won't say anything about it to the health authorities, because they're ashamed to say that they ate fruit from the trees after the accident."

Signora Brambilla went on to say that one of the reasons the people in the apartments weren't evacuated was that pressure was put on the municipal politicians by what she called "brass hats" among the cabinetmakers and craftsmen along the road near the block, who were afraid of losing business. She said that the people on the block had not been given proper medical attention, and that they had made many protests about the way the decontamination program had been carried out. When she and her neighbors demanded that their apartment house be properly cleaned and decontaminated, she said, "the municipality said that it wasn't their job, it was Spallino's, and Spallino said he'd given money to the municipality to have it done, but, of course, it isn't being done—nothing has happened." She said there had been so many contradictory actions taken by the regional authorities that it was no wonder a great many Seveso people had stopped considering dioxin dangerous. "The authorities allowed the people to go in and out of Zone A, and take out things like mattresses, and the people brought their dioxin out with them," she said. "And when people would scold them for taking their things out of the zone they'd turn around and say, 'You don't think you have dioxin in *your* house? You have dioxin, just as we do!' "

One of the women said that she had seen the way the decontamination teams cleaned one of the schools in the area—with Spic and Span as the decontaminant. "And in the gymnasium where they had stored dead animals for a time they repainted the walls but they left the contaminated fibre matting on the floor for the kids to continue doing gymnastics on." Only a month before, she said, she and some neighbors had spotted trucks moving along the road right by the apartment building. They were transporting contaminated earth and foliage by some circuitous route to be dumped in the most polluted part of Zone A. The trucks weren't covered, she said, and were spilling contaminated material out as they went. The women telephoned the authorities to protest, but the only result was that more trucks came, whereupon the women went out into the street themselves and stopped them and turned them back in the direction they had come from.

Speaking of the situation in local schools, Signora Brambilla pointed out that although schoolteachers were listened to when they said they were afraid to continue working very long in schools in the area, when the students at the high school in Cesano Maderno—including her own great-nephew—publicly demonstrated for regular health checks for themselves, they were surrounded by hundreds of police, and people were saying things like "Those kids are on drugs!" She asked, "Isn't it a scandalous thing that when schoolchildren come to ask for blood tests three hundred armed carabinieri are brought in to deal with them?"

I remarked to Signora Brambilla that I gathered she and her neighbors hadn't attended the arrival of Cardinal Colombo to bless the houses in Subzones A-6 and A-7.

Signora Brambilla said they hadn't. "The Church has done no more for us than the regional authorities," she said. "We in this place, we're regarded like smoke in the eyes."

Outside the Trattoria Savina, I got into a discussion with a youngish man in blue jeans, who told me that he had a small building business literally on the border between Zone A and the Zone of Respect, in the southern part of Seveso—his warehouse, which was attached to his house, was in Zone A, but the house itself was just over the line and in the Zone of Respect. His warehouse containing all his building materials had been pronounced contaminated with dioxin and had been sealed off. His business was in terrible shape, he said. Before the ICMESA explosion, his "order book was bursting." Now he was left with almost nothing. He drove me to his home. The garden outside was cluttered and had an untended look—he said they had been told by the authorities not to disturb it because of the dioxin levels there. He invited me into the living room of his house, which he told me he had built with his own hands, and, inside, he introduced me to his wife and a friend of his, a worker who lived nearby. The friend said that before the ICMESA accident he had been healthy, but that now he had liver trouble.

In the living room were the two small children of the young builder and his wife. I asked his wife if she was worried about the health situation arising out of the ICMESA accident. She said she was very much worried about it—and particularly about the future of their children. She said that because their house was in the Zone of Respect, no decontamination work on it had been even attempted by the authorities. The garden had to stay exactly as it was—those were the orders given them. But the children had to play somewhere, she said. During the previous summer, she and her husband had sent the children to Sicily, where they came from, for three months at their own expense, to get them away from the area and dioxin. They had been unable to obtain any compensation for their business and personal losses. They would like to sell their house, she said, but with things as they were nobody would offer much more than a quarter of its worth on the open market. Now it was winter, she said, and she had to try to keep the children in the house and out of the garden, which was difficult. "I don't even dress them for the outdoor weather," she told me. "I keep their overcoats separately from their other clothes, so that they really can't go outdoors for long in the cold weather. It seems to be the only way of keeping them away from harm."

I DROVE southeast to a point where Zone B bordered on the Zone of Respect. There were a number of stores along the street I was travelling on. Noticing a pharmacy on one corner, I parked and went inside. It was an antiseptic-looking shop, its counter uncluttered with displays of antacid remedies, nasal sprays, and the like. A few customers were at the counter. I waited until the pharmacist, a trim-looking man in a white coat, had served them, and then had a few words with him. I told him I was interested in the community's views on dioxin contamination and on the official programs for dealing with it.

The pharmacist shrugged and said, "Apart from time's

being the best and worst healer, one can say only that here, unfortunately, everything finishes in politics. This tragedy has turned into a farce. A farce of cash considerations. I suppose I'm a bit cynical about it all. What you have here is an insidious, odorless, invisible poison, and people react differently to the thought of it. Some people come in and say, 'The regional authorities say you can't grow carrots. I'm growing carrots and eating them, too!' And you have people who show that they are really frightened—but they can't do anything about it."

While the pharmacist was talking, two other customers had come up to the counter, and they joined in with comments of their own. The first was a middle-aged man. "Dioxin! Since the accident, I've passed along the superstrada through here six times a day in my car, and I'm not dead yet," he said.

The second customer, an elderly woman, said, "There never was a problem. We've been living with dioxin all along. Any pollution is called dioxin. All these factories have been spitting it out for thirty years."

Leaving the pharmacy, I drove on westward, to find my way onto the superstrada and head back to Milan. A block or so away from the pharmacy, I noticed a shoe store. I parked and went in, and asked a woman who appeared to be in charge if by any chance her store repaired shoes. She said that it did—that the repair work was done by her husband, who would be there in a few minutes. I introduced myself and told her that I wondered if doing shoe-repair work in this particular area had caused her husband any concern or any problems.

The woman received my remarks as though she had been waiting a long time to unburden her mind to someone on that very subject. In an animated voice, she told me that her husband, whose name was Leonardo Puglisi, had been repairing shoes for years as part of his shoe business. She said that some months after the explosion he had become worried that repairing shoes worn by people who were perhaps tracking dioxin around on the soles might be a hazardous thing to do on a regular basis. She said that between April and October of 1977, on his own initiative, her husband had stopped repairing shoes. "He wouldn't touch them anymore." At one point, he had gone to the municipal authorities in Seveso and asked them if it was safe to repair shoes from the area regularly. She said he was told that although it was true that dioxin could be carried around on the soles of shoes, there was no great problem, and he could carry on his repair work with safety, but that if he wanted to be particularly careful he could always give the shoes brought in to him for repair a washing off with water, or even a little alcohol. She went on, "When my husband stopped repairing shoes, his customers didn't react well at all. He told them he was worried about the possibility that there was dioxin on the soles of shoes he would have to repair. The customers were indignant. They said that was a lot of nonsense. They didn't even want to think about the problem. They said to him, 'All the other shoe repairers are doing it. Why not you?' They told him that they wouldn't buy new shoes from him if he wouldn't repair them when they got worn. They said they would go to the next town, where they could get their shoes repaired at the same store where they bought them. So my husband had to give in and start repairing shoes again. If he hadn't, I think we might have gone out of business. As it was, business had dropped off badly when he stopped repairing shoes. Even now, when my husband tells people that he's careful because of the dioxin they laugh at him."

As we were talking, her husband came into the store. Puglisi was a neatly dressed man of about forty, wearing his overcoat draped loosely over his shoulders in the Italian style. Accompanying him was his teen-age daughter, Maria, an attractive girl. Signora Puglisi told her husband who I was and what we had been discussing. Signor Puglisi told me that although he had so far suffered no ill health as a result of working with shoes, he continued to be worried about the long-term effects of this exposure. He said that after the accident his wife and her mother had both developed skin rashes and that several months later his daughter had developed kidney trouble. He showed me his small workroom, which was in a partitioned-off area to one side of the store. The workshop was immaculate; there were none of the piles of leather scraps and dust usually to be seen lying about in shoe-repair shops. All his tools were laid out in an orderly way. He told me that he was careful to keep his workshop swept up, and that he put all the worn soles from shoes to be resoled in plastic bags, which were picked up by the municipal garbage collectors. He just wished he could stop the repair work altogether. "If we could sell the business and move away from here, we would. But we know we can't sell, with conditions as uncertain as they are. And I can't just stop and close up shop. This is my trade; this is what I'm trained to do."

"Now that it seems that we must stay here, what can we do to live with dioxin?" Signora Puglisi said to me. "Isn't it useless for the authorities to send old people and children away to the sea for a month? What about the people who don't get away? And what about the eleven other months, with dioxin all around you? We're from Sicily, my husband and I. From Taormina. Oh, it's so beautiful there! If we could sell the place, we'd go down there immediately. Away from here."

4

IMPLICATIONS OF DIOXIN CONTAMINATION

THAT THE immediate effects of the ICMESA explosion on the people of Seveso and the surrounding area were not very much worse than the available information indicates they have been is, of course, a source of great relief. But the implications of this situation could be readily misunderstood, particularly by those who may have been impressed by arguments advanced by the chemical industry that the fundamental resilience of the human constitution allows modern populations to contend successfully with the effects of chemical pollutants lying at very low levels in the environment. In view of the mounting evidence—from laboratory studies, from repeated industrial accidents involving trichlorophenol production and from the known devastating effects of dioxin on animal life—of the potential dangers of dioxin to people, the relatively light scattering of ill-effects so far manifested among the Seveso population tends to illustrate not the lack of dioxin's toxicity to humans, but the mysterious—one might say devilishly capricious—manner in which it can strike and how little is as yet understood about this substance.

As for the teratogenic, or fetus-deforming, properties of dioxin, which have been repeatedly demonstrated in animal experiments, the lack of significant data about marked teratogenic effects of the ICMESA cloud on Seveso women does not necessarily demonstrate that dioxin is in no way implicated in human birth defects. One recent development that creates suspicion in this respect is reported difficulties associated with repeated use of hexachlorophene —the antibacterial compound for the manufacture of which the ICMESA factory was producing trichlorophenol when the reactor exploded. As I mentioned earlier, the Givaudan people themselves conceded, in 1977, that their hexachlorophene was commonly contaminated with dioxin at levels of up to twenty parts per billion. (They now say that that level has since been lowered to ten parts per billion.)

In June, 1978, in New York, Dr. Hildegard Halling, chief physician of the Department of Chronic Somatic Diseases at Södertälje Hospital, near Stockholm, presented a paper at a symposium on environmental health hazards held under the auspices of the New York Academy of Sciences, in which she set forth the results of a six-year study she carried out on the physical condition of infants born to nurses and other female medical personnel who worked in six hospitals in Sweden. As part of the hospital hygiene, these women had washed their hands in liquid soap containing hexachlorophene as many as sixty-seven times a day, and had done so well into their pregnancies. Dr. Halling found that twenty-five of the four hundred and sixty children born to these women were severely malformed and forty-six were born with minor deformities. In contrast, two hundred and thirty-three children born to women who did not use hexachlorophene and who made up a control group had no major deformities at all and only eight minor deformities. Of course, in addition to the obvious consideration that hexachlorophene per se could be the prime suspected teratogen, it is possible that the soap contained impurities other than dioxin, and that these could also be implicated in the high incidence of malformations.

In any event, the association between trichlorophenol production, dioxin contamination, and ultimate hexachlorophene products and these observed human birth defects does underline the potential hazards associated with a whole chemical stew of chlorinated phenolic products regularly found to be tainted with dioxin or dioxinlike substances. While no firm connection between the widespread use of the 2,4,5-T herbicide and birth defects among women in or around sprayed areas has yet, it seems, been satisfactorily demonstrated in the many suspicious incidents reported in this country and in South Vietnam, there have been repeated complaints by women in or near 2,4,5-T-treated areas in this country of ill effects of various kinds, including miscarriages that occurred soon after the women were exposed to drift from 2,4,5-T spray operations. In the spring of 1978, a letter was sent to the Environmental Protection Agency by eight women who lived in sparsely populated parts of Oregon and whose homes were in or near areas that timber companies and other outfits, under contract to the Forest Service, sprayed with a mixture of 2,4,5-T and 2,4-D. The women claimed to have suffered miscarriages within weeks of spray operations during the previous four years. As this is written, these claims are under investigation by the E.P.A., which by April, 1978, after years of delay and inactivity, was at last prevailed upon to consider the case against permitting the continued use of 2,4,5-T in this country.* Meanwhile, the manufacture and use of the herbicide continued.

During 1978, a number of Vietnam veterans came forward to say that during the nineteen-sixties they were exposed to Agent Orange in areas defoliated by the American military and that the exposure has led to various illnesses, including cancer. So far, no sufficiently hard evidence of such a specific cause-and-effect relationship has been

*See *Federal Register,* April 21, 1978: "Environmental Protection Agency Pesticide Programs. Rebuttable Presumption Against Registration and Continued Registration of Pesticide Products Containing 2,4,5-T."

presented, and such a link would be difficult to establish without an elaborate study. But nobody representing the United States government has yet undertaken a systematic follow-up study of American military personnel likely to have been most heavily exposed to 2,4,5-T in Vietnam—for example, the crew members who loaded thousands of drums of Agent Orange aboard the C-123 cargo planes used in the defoliation operations. It is certainly disquieting to contemplate the lack of any serious attempt to ascertain the condition of such people in the light of recent reports about an extraordinarily high death rate among workers who had been regularly engaged in 2,4,5-T spray operations along railroad rights-of-way in Sweden. Nor is it less disquieting to contemplate the fact that—as I have pointed out—not only has beef fat from cattle grazed on Western rangeland previously sprayed with 2,4,5-T been found to contain traces of dioxin of up to sixty parts per trillion—thus putting dioxin into the human food chain—but that samples of mother's milk taken in the Western part of the country have also been found to contain traces of dioxin. Considering that many of the surviving offspring of female experimental animals exposed to dioxin have been observed to sicken and die primarily from the poisonous effects of ingesting dioxin-contaminated mother's milk, the prospect of any long-range invasion of the human maternal milk supply through dioxin contamination is one that surely calls for the most stringent federal regulatory action.

Certainly, on these and other grounds, an immediate prohibition on the further manufacture and sale of the dioxin-contaminated herbicide 2,4,5-T seems to be an appropriate first order of business for regulatory bodies and legislators. Even so, an outright ban on all uses of 2,4,5-T as presently formulated—and, let us say, the application of most stringent regulations on an international scale concerning the manufacture and applications of hexachlorophene—would not in itself eliminate the dioxin problem. As I have tried to emphasize, in various articles on the subject in *The New Yorker* since the early 1970s,* dioxin and dioxinlike contaminants and their chemical precursors are present in an array of polychlorinated derivatives of chlorophenols, many of them associated with the production processes used to manufacture 2,4,5-T, and associated, particularly, with the 2,4,5-trichlorophenol that serves as the starting material for so many of these compounds. Commercial chlorinated phenolic products are commonly used as slime-killing agents in paper-pulp manufacture, and as stabilizers or fungicidal agents that are incorporated in a wide range of consumer products, including adhesives, water-based and oil-based paints, varnishes and lacquers, and paper and paper coatings. The annual world production of chlorinated phenolic compounds has been estimated to be about one hundred and fifty thousand tons. One such compound, which is a by-product of 2,4,5-T manufacture, is pentachlorophenol, a fungicidal preparation very widely used as a wood preservative. It has also been incorporated into paints, shampoos, and laundry starches. Many commercial pentachlorophenol products are known to be contaminated with dioxins or dioxin precursors of varying toxicity, and the history of such products is replete with highly unpleasant incidents connected with their use. For example—as I reported in 1970 in an article in *The New Yorker* that attempted to point to the potential hazards of various products associated with 2,4,5-T production—pentachlorophenol is known to have had fatal effects on human beings. A paper appearing in the *Journal of Pediatrics* in August, 1969, described the cases of nine infants between six and fourteen days old who were born in St. Louis, and who were subsequently taken to St. Louis

*See Bibliography, page 200.

Children's Hospital with a severe form of an undiagnosed illness. The symptoms of this illness were excessive sweating, increased heart rate, respiratory difficulty, and liver enlargement. Shortly after being admitted to Children's Hospital, two of the infants died; the others survived after being given blood transfusions and other treatment. Eventually, the cause of the illnesses was traced to an antimicrobial laundry product that had been used in large amounts in the laundry of the hospital where the children were born. This product contained sodium pentachlorophenate, and traces of pentachlorophenol that had remained in diapers and other clothing after they had been laundered had penetrated the children's skins and entered their body systems. Such is the toxic nature of pentachlorophenol that even after the rinse was discontinued traces of pentachlorophenol continued to be found in the blood of newborn children and of expectant mothers at the hospital. It turned out that the expectant mothers had been lying between sheets previously laundered with the pentachlorophenol rinse, and it appears that even with this supposedly slight exposure the pentachlorophenol had penetrated into the mothers' bodies and crossed the placental barrier, entering the systems of the unborn infants. This reported incident reminds one in ominous fashion of the fate of the infants of the Swedish nurses who, over the years, had regularly washed their hands with hexachlorophene preparations, as reported in Dr. Halling's study.

A further illustration of the toxic potential of pentachlorophenol is the loss of entire dairy herds in the area of Kalkaska, Michigan, in 1976 and 1977. During those years, more than four hundred cows died of poisoning diagnosed as having been caused by pentachlorophenol. In the most severely affected herd, sixty-one of the sixty-seven calves born to surviving cows were stillborn, and many were deformed. Eventually, it was discovered that wood in the recently built barns in which the disastrously affected herds had been kept had been previously treated with pentachlorophenol. The cows evidently had rubbed against the treated wood, and probably had licked the surfaces of their wooden feeders, which also had been treated with pentachlorophenol. Pentachlorophenol levels in the dead animals ranged as high as 1,136 parts per billion in the blood and up to thirteen hundred parts per billion in the liver.

In this country alone, the annual production of pentachlorophenol is about sixty million pounds. One of the particularly important implications of the very widespread use of pentachlorophenol in wood is that when samples of commercial pentachlorophenol formulations have been burned, this burning process has shown the capacity not only to release a whole array of dioxins and dioxin precursors of varying toxicity, but also actively *to create* further amounts of these undesirable compounds. Since the fate of most wood that does not eventually rot away is disposal by burning, this ever-continuing process in itself could be responsible for releasing large amounts of various dioxins and dioxin-like substances into the atmosphere and environment over the years.

And the list of other potential routes of dioxin contamination is growing. Scientific papers by Drs. Hans Rudolf Buser and Hans-Paul Bosshardt, of the Swiss Federal Research Station, and Dr. Christopher Rappe, of the University of Umeä, Sweden, have reported the discovery of significant traces of polychlorinated dibenzodioxins in the fly ash from a municipal incinerator and an industrial heating facility in Switzerland, which were burning mainly household and industrial refuse. The levels of the various dioxins found in the fly ash ranged up to six-tenths of one part per million, and, while TCDD itself was a minor component of the contaminated material, another highly toxic dibenzodioxin, of the hexa variety, was a major component in the fly ash taken from the industrial facility. Interestingly, the authors have suggested, on the basis of further laboratory tests, that the dioxins discovered in the incinerator fly ash might have originated in commercial chlorinated phenolic products presumably present in the incinerated material.

There are yet other roots. Dioxin-contaminated wastes from a number of other factories that have produced 2,4,5-trichlorophenol certainly have been dumped over the years, along with other toxic wastes, in landfills scattered around the country. These wastes constitute chemical time bombs capable of exerting extremely harmful long-range effects on nearby communities. To cite one such case: For a number of years the Givaudan corporation in the United States was supplied with 2,4,5-trichlorophenol for its manufacture of hexachlorophene by the Hooker Chemicals and Plastics Corporation, of Niagara Falls, New York. As this book was prepared for press, evidence was coming to light that many of the toxic wastes from Hooker's trichlorophenol production in the 1940s and early 1950s had been buried, along with other toxic debris from the Hooker Niagara Falls facility, in a Hooker-owned landfill in Niagara Falls known as the Love Canal. In 1952, this site was partly filled up and sold by Hooker to the Niagara Falls Board of Education for a dollar; the site and surrounding terrain then became a built-up residential area. Persistent reports of noxious odors and of suspicious substances on the ground led to the discovery late in 1977 that toxic chemicals buried in the dump were leaching out of the soil at the site. Subsequently, analysis of samples taken from the site showed the presence in the buried material of some two hundred chemical compounds, many of them highly toxic. By that time, complaints of ill health among local residents were being reported in the local press. By November, 1978, Hooker officials conceded that perhaps two hundred tons of trichlorophenol wastes also might have been included in the chemical debris buried in the Love Canal dump. Shortly after this admission, high concentrations of trichlorophenol were in fact detected in effluent from the site, and by December traces of TCDD itself were measured in leachate from the Love Canal. On the basis of measurements made of trichlorophenol wastes involved in the 1971 Missouri dioxin-poisoning episode previously referred to, one may calculate that the amount of TCDD contaminant within wastes buried at the Love Canal site could total about sixty kilograms, or about a hundred and thirty pounds—many times more than the total amount of dioxin reckoned to have been released as a result of the Seveso explosion, and an amount that may equal the estimated total dioxin content of the thousands of tons of Agent Orange sprayed over Vietnam in the years of American herbicidal warfare operations there. And other Hooker chemical landfills not far from the Love Canal site in Niagara Falls could turn out to contain further large amounts of TCDD in trichlorophenol wastes probably buried at these sites.

All of this mounting evidence reinforces, then, the prospect of a seemingly persistent and accumulative burden of dioxin contamination imposed on populations throughout the industrialized world. The wider one casts the net of inquiry, the more revealing appears to be the evidence indicting not only TCDD and other dioxins but an array of structurally similar substances that fall within the general chemical designation of polyhalogenated aromatic compounds. This group includes such substances as polyhalogenated benzenes and toluenes, biphenyls (PCBs), terphenyls, naphthalenes and diphenyl ethers, which may contain such potent dioxinlike contaminants as dibenzofurans. A number of such compounds, some of which are widely used as flame retardants, have been shown, again and again, to exert extensive malign effects on man and animals by their toxicity and by their accumulative propensities in the environment and in living tissue. Toxicological and epidemiological accounts of their activity are permeated with disconcertingly similar symptoms and effects only too familiar to those who have followed the history of dioxin contamination: chloracne-like manifestations, liver damage, nervous-system and vision impairments, spontaneous abortions and birth defects, abnormal functioning and derangement of immune systems, and carcinogenic potential. And there is one other similarity: there is no known antidote to these increasingly ubiquitous toxins.

The issue of dioxin and the 2,4,5-T herbicide stands, it seems to me, as a sort of regulatory touchstone concerning the appropriate and permanent measures of control that clearly need to be undertaken to cut or eliminate the flow of such toxic substances into our environment. The dioxin issue characterizes the broader problem of bringing under control the introduction into the environment, often at admittedly very low levels, of contaminants that are nonetheless of extraordinary toxicity and persistence. And this is why the chemical industry has fought over the 2,4,5-T issue with such energy and assiduity—in the courts, in regulatory proceedings, and in its public-relations activities. A defeat for the chemical industry over the issue of halting the production of 2,4,5-T must entail the danger of a regulatory precedent for controlling not only the slow environmental leak of dioxin, but also of a whole group of associated compounds having markedly toxic, persistent, and accumulative properties. Up to the present, many such compounds besides 2,4,5-T have escaped effective regulation because the chemical companies producing them have managed, through the frailties of existing law and the carelessness of federal regulators, to have the sale of their products in interstate commerce certified on a product-by-product, formulation-by-formulation basis, to the exclusion of broader considerations of public hazard. In their dealings with regulatory agencies, the companies have usually been able to obtain the registrations necessary for the distribution of the products by making much of the very low levels at which the toxic elements of a particular formulation are likely, when spread out, to contaminate man and the environment. But, unfortunately, the human body does not discriminate between, let us say, TCDD as laid down by Dow Chemical's herbicidal sprays, and TCDD laid down by the effluent of municipal incinerators that might be burning the pentachlorophenol formulations of another chemical company. Surely, what proper regulation should protect the public from is not merely the effects of particular product-brand formulations but the collective and accumulative effects over the long term of whole classes of toxic compounds, particularly those containing contaminants of similar persistence. Yet, as this book is completed, there has been no fundamental federal regulatory change on the 2,4,5-T issue since 1970. But, like the invisible dioxin that continues to haunt the human environment into which this poison has been released over so many years, the issue will not go away. Whatever may happen concerning the federal regulation of 2,4,5-T and dioxin, the 2,4,5-T issue, as I have tried to emphasize in my writings on this subject, is particularly important in that it touches on and symbolizes the encroachments of heedless technology upon the fundamental liberties of the individual, and, above all, touches upon our guardianship of the physical integrity of those yet unborn.

APPENDIXES

TCDD and Industrial Accidents

Robert W. Baughman

Some of the halogenated dibenzo-*p*-dioxins are extraordinarily toxic. All of the non-halogenated derivatives that has been tested are relatively non-toxic. The toxicity of the halogenated derivatives was discovered accidentally through the appearance of chloracne and other severe toxic response in workers involved in the synthesis of chlorophenols and their derivatives, and through the death of large numbers of chickens from pericardial edema and other lesions caused by a toxic fat food supplement containing chlorinated dioxins.

CHLORACNE

The term chloracne was used first in 1899 by Herxheimer[118] to describe a severe form of acne observed in workers in plants producing chlorine by electrolysis. The disease was characterized by numerous comedones and retention cysts which later were shown to be caused by hyperkeratosis. Herxheimer attributed the disease to exposure to chlorine gas, but in retrospect it seems likely that the actual cause was chlorinated aromatic compounds formed by the action of the chlorine on tar used to protect the electrolysis towers.[119] Since that time chloracne outbreaks have oc-

---

Excerpt from "Tetrachlorodibenzo-*p*-dioxins in the Environment: High Resolution Mass Spectrometry at the Picogram Level." Ph.D. dissertation, Harvard University, 1974. Reprinted by permission.

curred periodically in chemical industries involved in the synthesis of chlorinated aromatic compounds. In the early part of this century several incidents were reported with chlorinated naphthalenes and biphenyls which were used widely as insulators in the electrical and radio industries.[119,120] Other pathological symptoms were reported, in particular, damage to the liver.[119,120] Because of the involvement of chlorinated naphthalenes, Wauer and Teleky[121,122,123] suggested that the term chloracne be replaced by pernakrankheit or perna disease derived from *per*chlorinated *na*phthalene. In the United States during the second world war chloracne was observed on a large scale when large amounts of chlorinated napthalenes were used to insulate electrical cables which protected ships against magnetic mines.[119,120] Several fatalities involving liver necrosis were reported.[119] Shelley and Kligman[124] showed that penta- and hexachloronapthalenes were stronger chloracnegens than were either the higher or lower chlorinated homologs. They reported that in conjunction with chloracne in man, "Many cases of fatal hepatic necrosis have been observed." In 1947 Olafson[125] reported a disease in cattle, termed X-disease, that was characterized by hyperkeratosis, liver degeneration and other symptoms. A rapid decline in vitamin A levels was observed which it has been suggested may have been the result of the effects on the liver.[126a] The cause of the disease was traced to animal feeds contaminated with chloronapthalenes.

By the 1950's use of chlorinated napthalenes had dropped considerably, but a new and even more potent source of chloracne was encountered in the production and processing of chlorophenols. This new source of chloracne in fact had made its presence known during the 1930's when chlorophenols were introduced by the Dow Chemical Company under the trade name of Dowicides, as biocides, in particular as wood preservatives.* Patents were obtained in 1935 and 1936 for the use of alkali metal salts of 2,4,5-trichlorophenol as fungicides.[126b,126c] The Dowicides were described in a company prospectus dated 16 December 1936.[126d] However, even before this prospectus was published, the first reports of severe chloracne associated with use of the chlorophenols had reached the medical literature.

Earlier that same year severe cases of chloracne were reported in lumber workers in Mississippi involved in treating wood with a fungicidal chlorophenol formulation, apparently Dowicide H, which consisted of primarily tetrachlorophenol.[127] In a more detailed report, Stingily[128] indicated that three or four hundred persons were involved. Initially erythema and ulceration of the skin were observed, followed by formation of comedones, cysts and pustules, thickening of the skin and urinary disturbances. In some cases leg cramps, thrombosis or highly colored urine were present. The lesions were reported to be remarkably persistent, sometimes lasting several years after the final exposure. Apparently the chickens soon came home to roost as in 1937 Butler[129,130] reported a similar incident at the Dow Chemical plant in Midland, Michigan involving twenty-one workers who had handled 2- (2-chlorophenyl) phenol and tetrachlorophenol. Marked hyperkeratosis was

---

*In a way the chlorophenols are a geminal product of Dow Chemical Company. The company founder, Herbert Henry Dow, made his start in the chemical industry in the 1890's by extracting bromine and chlorine from brine.[126e] In 1918 he obtained a patent for the preparation of phenol by hydrolysis of bromo-benzene.[126f] This procedure was adapted to hydrolysis of chlorobenzene, and by 1930 the company was operating the largest synthetic phenol plant in the world.[126g] The marriage of the basic chemical products chlorine and phenol soon produced as offspring the chlorophenols and, unwittingly, the chlorinated dibenzo-*p*-dioxins.

observed with "enormous numbers of comedones . . . in some cases so numerous as to produce a black discoloration . . . ." Severe scarring and persistence of the condition were noted. In his paper Butler stated that these chemicals should not be used as biocides until the mechanism by which they caused chloracne was understood. The plant was closed temporarily.

Butler indicated that experimentation with animals would be undertaken to attempt to elucidate the mechanism of toxicity, but he was not supported in this effort by the Company.[130] If these experiments had been completed, it is possible that the chlorodioxins would have been identified as being highly toxic twenty years before the eventual discovery in 1957.

In 1941 Adams *et al.*[131] of Dow Chemical reported a bioassay for chloracnegens which consisted of applying suspected agents to the interior surface of the rabbit ear and monitoring acne-like changes. They tested five types of chemical products which were acnegenic, including crude chlorophenols. Although they did not discuss which component of each product may have caused the chloracne, the fact that of the classes of compound they examined the term "crude" was attached only to chlorophenols suggests that they may have been aware that in this case compounds other than the main chlorophenol products were involved.

Following the discovery of the herbicidal properties 2,4,5-T and related phenoxyacetic acids during the second World War,[132] production of this compound and its precursor, 2,4,5-trichlorophenol which was prepared by hydrolysis of 1,2,4,5-tetrachlorobenzene, began on a large scale. This production was accompanied by a series of new outbreaks of chloracne. In 1949, 228 workers developed chloracne as the result of an accident in Nitro, West Virginia at a 2,4,5-T plant owned by the Monsanto Chemical Co.[133a,133b] Other symptoms observed included severe pains in skeletal muscles, shortness of breath, intolerance to cold, palpable and tender liver, loss of sensation in the extremities, demyelination of peripheral nerves, fatigue, nervousness, irritability, insomnia, loss of libido, and vertigo. Porphyria was not reported. In 1951 Baader and Bauer[134] reported chloracne in 17 workers in Germany who had participated in an investigation of industrial syntheses of 2,4,5-trichlorophenol and pentachlorophenol.

Within a few years similar incidents were reported at other 2,4,5-T and 2,4,5-trichlorophenol plants.[135–140] In 1956 Dugois and Colomb[138] described one such case involving 17 workers at a 2,4,5-trichlorophenol plant near Grenoble, France. In a unique observation they reported that in addition to having severe chloracne and internal disturbances, those affected gave off an *"intense odeur chlorée."* They suggested that the source of the disorder was some chlorinated cyclic hydrocarbon such as a chloronapthalene. In 1957 Hofmann[135] described in detail an incident that had happened in November 1953 in Ludwigshafen am Rhein at a 2,4,5-T factory owned by Badischer Anilin- & Soda-Fabrik. Hofmann reported that he had found a likely candidate for the toxic agent. He established that the unknown compound was neither 2,4,5-trichlorophenol nor its precursor 1,2,4,5-tetrachlorobenzene and that it must be at least a hundred times more toxic to rabbits than the chloronapthalenes. In fact, the compound was so toxic that control animals in cages next to treated animals developed liver necrosis as did untreated animals which were placed in cages previously housing treated animals. This behavior at first led Hofmann to believe that they were dealing with a virus infection. In retrospect, it appears that cross contamination may have occurred via the excreta.

Various chlorinated aromatic ethers and diphenyls were investigated and found not to be particularly toxic. Then a number of chlorinated dibenzofurans were synthesized. Those containing three to five chlorines came close to having the required level of toxicity, both in terms of hepatotoxicity and chloracne. Hofmann proposed that these dibenzofurans were the source of the problem in the 2,4,5-T and trichlorophenol plants.

At the same time in Hamburg, Schulz and Kimmig[136,137] were carrying out an almost identical investigation. They too found that certain chlorinated dibenzofurans were extremely toxic. However, as Schulz[136] pointed out, these compounds did not seem to be present in the technical chlorophenols. The solution to the problem appeared in a most ironic way. A man with an unusually severe case of chloracne, who had no connection at all with the 2,4,5-T or chlorophenol plants, was referred to Schulz. This man was the assistant working with Sandermann (see above) in the attempt to confirm the structure of Merz and Weith's original "perchlorophenyleneoxide" by chlorinating dibenzo-*p*-dioxin to octachlorodibenzo-*p*-dioxin.[136] Because of decreased solubility and decreased reactivity the reaction had stopped at 2,3,7,8-tetrachlorodibenzo-*p*-dioxin (TCDD). Kimmig and Schulz readily confirmed the extraordinary chloroacnegenic and hepatotoxic properties of this compound in tests with rabbits.[137] In a self-administered skin test, Schulz demonstrated that TCDD is chloracnegenic in humans.[142a] Schulz suggested that chloracne also might be produced following ingestion of TCDD, but this was not tested.

At about the same time, a similarly inauspicious episode in which the toxic agent was *not* recognized, occurred in the United States. As a part of an investigation of the substitution reactions of dibenzo-*p*-dioxin, Dietrich synthesized various halogenated derivatives, including the 2,3,7,8-compounds. He subsequently became severely ill and was admitted to the cancer ward of the Atomic Energy

Commission Medical Center in Chicago. Fortunately, after an extended period of hospitalization he slowly recovered, although the chloracne remained active for about five years. In the case of Dietrich's misadventure the connection with the compounds he was synthesizing was not made. Only several years later in 1965, at about the time a new series of chloracne incidents struck 2,4,5-T plants in the United States (see below) did he learn the cause of his illness. Dietrich's story provides another note of irony. After he had synthesized various chlorinated dioxins, they were submitted to the Army for medicinal evaluation. The only compound (a non-halogenated one) which was selected for medicinal use was used in a dermatological salve.[141]

Having shown that TCDD was an extraordinarily potent chloracnegen, Schulz and his colleagues then demonstrated that this compound was formed during the synthesis of 2,4,5-trichlorophenol from 1,2,4,5-tetrachlorobenzene. The conditions required to hydrolyze the tetrachlorobenzene to trichlorophenol were found to be sufficient to cause a small fraction of the trichlorophenol to condense to form TCDD according to synthetic Route A of the preceding section. After the nature and identity of the toxic material were established, conditions (unspecified) were found which reduced the level of dioxin in the product and which reduced the exposure of the workers to whatever dioxin was present. This resulted in a substantial reduction in the incidence of chloracne.[142a]

Schulz and his collaborators went on to carry out an extensive investigation of the symptomology and etiology of chloracne. In summary the symptoms were described as follows:[142a]

> Numerous comedones formed, first on the face, especially on the cheeks above the malar bones, forehead, temples, chin and ears, after which folliculitis,

> pustules, boils and retention cysts occurred as a result of secondary infections. As the disease progressed, these symptoms spread in the majority of patients, especially to the sides of the neck, back of the neck, upper half of the back, chest, forearms, genitals and thighs. Numerous boils formed, particularly on the back of the neck and on the back. The efflorescences were generally located so closely together that scarcely any follicles remained unchanged.

Symptoms also included hyperpigmentation, erythema, swelling of the skin, hirsutism, loss of appetite, loss of weight, gastritis, severe liver damage, pulmonary emphysema, dyspnea, myocardiac damage, edema, renal damage, plus a long list of neurological and psychopathological symptoms including muscular disturbances, disturbances in memory and concentration, decrease in initiative and interests, depressions, disturbances in libido and potency and weakness in mental capacity. The dermatological symptoms were reported to be extremely obstinate, in some cases lasting for many years.[142a]

The number of workers involved or their individual symptoms were not given. In referring to this incident, May[142b] reported that some fatalities occurred as a result of severe liver damage, and that fifteen years after initial exposure another individual was expected to die soon of this cause. One death from pulmonary carcinoma was rejected as being attributable to dioxin exposure, while one from intestinal sarcoma was accepted. This last observation may be significant in view of the extensive possibly pre-cancerous lesions observed in the intestinal tract of monkeys and other test animals fed a diet containing chlorinated dioxins.[142c]

Recently Goldmann[143a] updated and expanded the description of the symptoms of the 42 workers who had been exposed to dioxin following an accident at the BASF plant at Ludwigshafen am Rhein in November 1953[135] (see above). Especially valuable are detailed individual case histories for periods up to eighteen years after exposure. A few of these case histories will be summarized here to supplement the generalized description that has been given thus far of the response in humans to TCDD.

One case involved a man who had never entered the building in which the accident occurred. He merely sat next to exposed workers in the lunchroom. Chloracne developed on the face and forearms. No other symptoms were observed.

In another case chloracne developed which was characterized by hen's egg-sized abscesses. In this individual comedones and retention cysts were still occurring eighteen years after exposure.

A third case involved a mechanic who worked for three days in the area of the autoclave used to hydrolyze tetrachlorobenzene to trichlorophenol. On the second day pains in the spine and headache occurred. Inflammation and swelling of the skin was followed by chloracne. Numerous pustular infections occurred which were not improved by treatments with antibiotics or vaccines. Myocardiac damage, toxic nephrosis, massive bronchitis, sinusitus, tonsillitis, and liver damage were observed. The effects on the liver included subacute hepatitis with formation of groups of hyaline bodies and fat vacuoles and hypertrophy. Thrombophlebitis of the left leg developed. The patient never regained his health. Ten years after the original exposure he succumbed to coronary insufficiency and lung embolism with extensive hemorrhaging.

A fourth case involved a 57 year old employee who worked in one of the autoclaves in 1958, five years after the original accident. Over this period of time the autoclaves were never used to prepare trichlorophenol. Although he was wearing a complete protective suit, he had removed his mask several times to wipe perspiration from his face. Within four days headaches, loss of hearing, and chloracne were present. One month later he was hospitalized with angina pectoris. Six months later acute pancreatitis developed and a large and very painful tumor was found in the upper abdomen. Death attributed to acute pancreatitis occurred shortly thereafter. Autopsy revealed intestinal ulceration and liver and adipose necrosis.

A fifth case involved a young bricklayer who spent two hours in the autoclave room repairing a wall. A persistent and severe case of chloracne developed. After about a year had passed his temperature became noticeably elevated and a massive X-ray-opaque area appeared in the left lung. Severe hemorrhagic pleuritis followed. Animal tests and cultures were negative. Eventually his condition began to improve slowly. Then about four years later he suffered acute psychosis with insomnia, loss of affect, hearing of voices, suicidal tendencies, physical discomfort and a burning sensation in the back. Within a short time he committed suicide by hanging.

Several general points can be made on the basis of the cases described by Goldmann. A very short exposure time can be sufficient to produce serious toxicity. Effects from a single exposure may last many years. Dioxin residues can persist in the chemical plant environment for at least five years. In the present instance the plant was sealed off after the incident in 1958 and later carefully demolished. One of the most prevalent symptoms appears to be decreased resistance to infection. Infections of the respiratory tract such as laryngitis, tracheitis, tonsillitis, and bronchitis were common. Neurological effects including loss of hearing, smell, and taste, polyneuritis and sensory and motor defects, and encephalomyelitis—a disturbance of the central nervous system with paralysis and spasticity on one side of the body—were observed in seven of the forty-two affected workers. Neurological damage appeared to affect the left side more than the right side. Drowsiness was a common complaint. Hemorrhages in various tissues and gastrointestinal disturbances, including ulcers, were often present. Many of these effects, and also effects on the heart, liver, and other organ systems, parallel those described by Bauer, *et al.*[142a] Porphyria was not mentioned. Except for the neurological effects, the symptoms observed in the workers are generally consistent with those observed in toxicity tests with TCDD in various animal species (see Mechanisms of Toxicity). (Few significant neurological effects have been observed in animals.) Perhaps the most striking aspect of the cases described by Goldmann is the tremendous difference in the nature and intensity of the responses from individual to individual. Differences in individual susceptibility also exist in animals, so much so that at times such differences have caused difficulty in determining LD_{50} doses for a given species.[143b]

The experiences in Germany described above were repeated on a substantial scale in the United States in the mid-1960's. In response to the greatly increased demand for 2,4,5-T which resulted from the U.S. military herbicide program in Vietnam, facilities and schedules were put under great pressure in an effort to increase production. Several outbreaks of chloracne occurred. In 1964, about the time the major phase of the herbicide program was getting underway, one such incident involving over seventy workers occurred at a plant operated by the Dow Chemical Company. An investigation was conducted which not surprisingly led to the conclusion that TCDD was the source of the problem.[133a] Although these results were not published, they were communicated to other manufacturers,

REFERENCES

118. K. Herxheimer, *Munch. Med. Woch., 46.1,* 278(1899).
119. K. H. Schultz, *Arbeitsmedizin-Sozialmedizin-Arbeitshygiene, 3,* 25(1968).
120. K. D. Crow, *Trans. of the St. John's Hospital Derm. Soc., 56,* 79(1970).
121. K. Wauer, *Zentralblatt für Gewerbehygiene, 6,* 100(1918).
122. L. Teleky, *Klinische Wochenschrift, 6.1,* 845(1927).
123. L. Teleky, *ibid., 6.1,* 897(1927).
124. W. Shelley and A. Kligman, *Arch. Derm., 75,* 689 (1957).
125. P. Olafson, *Cornell Vet., 37,* 279(1947).
126. (a) R. D. Kimbrough, *Arch. Environ. Health, 25,* 125(1972).
126. (b) L. E. Mills (to Dow Chemical Co.,), U.S. Patent 1,991,329(1935).
126. (c) L. E. Mills (to Dow Chemical Co.,), U.S. Patent 2,039,434(1936).
126. (d) Dow Chemical Co., Prospectus, 16 December 1936.
126. (e) W. Haynes, *American Chemical Industry, 6,* Van Nostrand Co., New York, 1949, p. 114.
126. (f) H. H. Dow, U.S. Patent, 1,274,394(1918).
126. (g) M. E. Putnam, *Plastics and Molded Products, 7,* 255(1931).
127. Queries and Minor Notes, *J. Amer. Med. Assoc., 106,* 2092(1936).
128. K. O. Stingily, *Southern Med. J., 33,* 1268(1940).
129. M. G. Butler, *Arch. Derm. Syph., 35,* 251(1937).
130. M. G. Butler, personal communication.
131. E. M. Adams, D. D. Irish, H. C. Spencer, and U. K. Rowe, *Ind. Med., 10,* 1(1941).
132. W. B. House, L. H. Goodson, H. M. Gadberry, K. W. Dockter, "Assessment of Ecological Effects of Extensive or Reported Use of Herbicides," Midwest Research Institute, Kansas City, Missouri, 1967, pp. 108–110.
133. (a) "Report on 2,4,5-T," A report of the Panel on Herbicides of the President's Science Advisory Committee, Office of Science and Technology, March 1971, p. 48.
133. (b) R. Suskind, "Chloracne and associated problems,"

Conference on Dibenzo-*p*-dioxins and dibenzofurans, Research Triangle Park, N.C., 2 April 1973.

134. E. W. Baader and H. J. Bauer, *Ind. Med. and Surg., 20,* 286(1951).

135. H. T. Hofmann, *Arch. Exp. Pathol. Pharakd., 232,* 228(1957).

136. K. H. Schulz, *Arch. Klin. Exp. Derm., 206,* 589(1957).

137. J. Kimmig and K. H. Schulz, *Dermatologica, 115,* 540(1957).

138. P. Dugois and L. Colomb, *Bull. Soc. Fr. Derm. Syph., 63,* 262(1956).

139. P. Dugois and L. Colomb, *J. Med. Lyon, 38,* 899(1957).

140. P. Dugois, J. Marechal and L. Colomb, *Arch. Maladies Prof., 19,* 626(1958).

141. J. J. Dietrich, personal communication.

142. (a) H. Bauer, K. H. Schulz, and U. Spiegelberg, *Arch. Gewerbepath. Gewerbehyg., 18,* 538(1961).

142. (b) G. May, *Brit. J. Ind. Med., 30,* 276(1973).

142. (c) D. H. Norback and J. R. Allen, *Environ. Health Perspec.,* (5), 137(1973).

143. (a) P. J. Goldmann, *Arbeitsmed. Socialmed. Arbeitshyg., 7,* 12(1972).

143. (b) J. B. Greig, G. Jones, W. H. Butler, and J. M. Barnes, *Fd. Cosmet. Toxicol., 11,* 585(1973).

144. J. Bleiberg, J. Wallen, R. Brodkin and I. L. Appelbaum, *Arch. Derm., 89,* 793(1964).

145. A. P. Poland, D. Smith, G. Metter and P. Possick, *Arch. Environ. Health, 22,* 316(1971).

146. M. F. Hoffman and C. L. Meneghini, *G. Ital. Derm., 103,* 427(1962).

147. N. E. Jensen, *Proc. Roy. Soc. Med., 65,* 687(1972).

148. K. A. Telegina and L. I. Bikbulatova, *Vestr. Dermatol. Venerol., 44*(3), 35(1970).

149. W. Braun, *Chlorakne,* Editio Cantor, Aulendorf, W. Germany, 1955.

The Evaluation of Possible Health Hazards from TCDD in the Environment

Matthew Meselson, Patrick O'Keefe, and Robert Baughman

For several years we have been developing and applying methods for the measurement of TCDD (2,3,7,8-tetrachlorodibenzo-*p*-dioxin) in the environment (1,2,3). TCDD is present as a contaminant in certain pesticides, including the herbicides 2,4,5-T and Silvex (4). Although the concentration of TCDD in these chemicals is very low, the great toxicity of TCDD and its possible accumulation in the environment make it advisable to determine how much TCDD is reaching various human populations and what exposure level might reasonably be considered hazardous to man.

The analysis of animal fat and milk is of particular interest because TCDD concentrates preferentially in lipid components of the body. Our current method for determining TCDD in fat and milk uses neutral extraction, four steps of column chromatography, and analysis by high resolution mass spectrometry (3). Before extraction, a known amount of the ^{37}Cl heavy isotopic isomer of TCDD that we synthesized for this purpose is added to each sample to serve as an internal standard. The great specificity and sensitivity of high resolution mass spectrometry make it especially well suited to the measurement of low levels of

Paper for Symposium on the Use of Herbicides in Forestry, Arlington, Virginia, Feb. 21–22, 1978. Reprinted by permission.

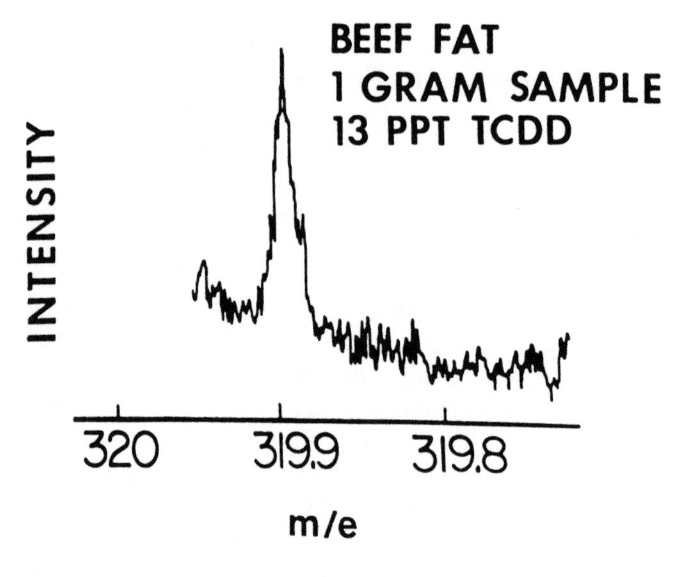

Fig. 1

TCDD. Figures 1 and 2 show examples of TCDD peaks as they are recorded by the mass spectrometer at the two TCDD mass/charge ratios which we routinely use for analysis, m/e = 319.897 and m/e = 321.894. In an individual mass spectrometer run, the amount of TCDD is determined by measuring the height of one or the other of these peaks relative to the height of the peak from the internal standard at m/e = 327.885 (not shown).

Figures 3 and 4 show the results of analyzing samples of beef fat and human milk containing various amounts of added TCDD, submitted to us by the Environmental Protection Agency and the Food and Drug Administration in order to test the sensitivity and accuracy of the analytical method. No TCDD above the limit of detection imposed

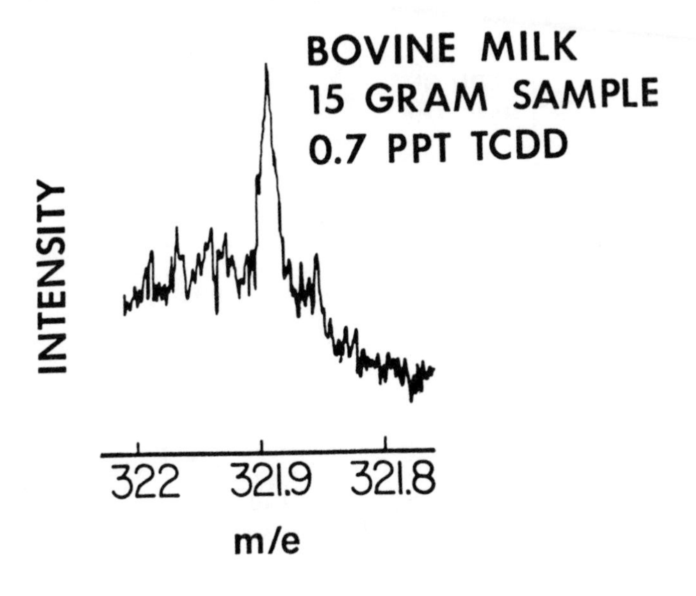

Fig. 2

by background noise in the mass spectrometer was found in control samples without added TCDD. As may be seen, the relation between added and measured TCDD levels is very close to linear over the entire range tested. TCDD was detected when added at levels as low as 2 parts per trillion (ppt) in beef fat and 0.25 ppt in human milk. However, near these limits, the measured amount of TCDD exceeded the amount added by a factor of up to three, an effect we are presently examining. Although analytical methods for TCDD have improved enormously over the last several years, further refinements are underway to permit accurate measurements at even lower concentrations and to provide improved discrimination among the positional isomers of TCDD, some of which may be present

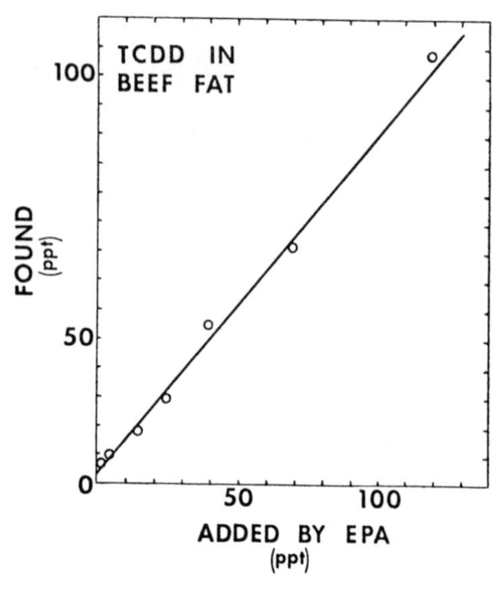

Fig. 3

in the environment in addition to the 2,3,7,8 isomer (5,6).

As part of the initial phase of an effort by EPA to monitor TCDD, analyses have been done by a number of laboratories on fat from cattle grazed on 2,4,5-T treated rangeland in Kansas, Missouri, Oklahoma, and Texas and from cattle grazed on untreated land. We received for analysis by our current method 14 samples from the 2,4,5-T group and one control. The samples were selected to include several which had been reported to contain TCDD by other laboratories.

We found TCDD in 11 of the samples from treated rangeland but none in the control or in beef fat samples from a Cambridge, Massachusetts market. The four samples with the highest levels were found to have 70, 24,

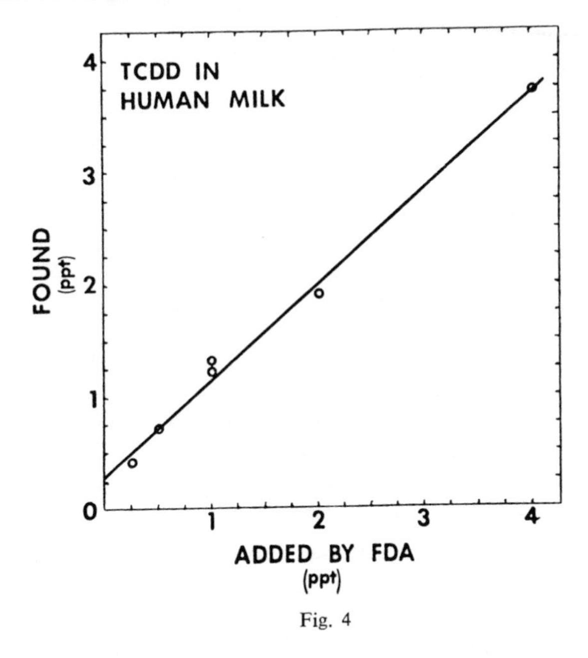

Fig. 4

20 and 12 ppt, respectively. The overall results of our analyses and those of others participating in the study were summarized by EPA in June 1976, as follows:

> Of the fat samples (85) analyzed, one shows a positive TCDD level at 60 ppt; two samples appear to have TCDD levels at 20 ppt; five may have TCDD levels which range from 5-10 ppt. While several laboratories detected levels (5-10 ppt) in this range, the values reported were very near the sample limits of detection. There exists a great deal of uncertainty of the analytical procedure below 10 ppt.

This interim summary needs a little clarification. Actually, the number of beef fat samples was 89, of which 68 were from the 2,4,5-T group and 21 were controls from unsprayed land. No consistent finding of TCDD was reported for the controls, of which 17 were analyzed at a sensitivity of 10 ppt or better, ten of them by more than one laboratory. Only 25 samples from the 2,4,5-T group were analyzed at a sensitivity of 10 ppt or better by more than one laboratory. Among these 25, there were 9 samples for which two or more laboratories reported positive TCDD levels, one sample at *ca.* 65 ppt, two at *ca.* 20 ppt and six in the range of 5-20 ppt. This ignores positive results obtained by low resolution mass spectrometry since they are unreliable. If one employs somewhat less stringent criteria for including samples in the tally, while still excluding low resolution positives, there are several more samples for which TCDD levels of *ca.* 5-30 ppt were reported, plus numerous ones in which TCDD was not detected. Since June 1976, EPA has accumulated more data and it is to be hoped that this and the data on which the 1976 statement was based will be released before much longer.

There appears to be a significant association between the use of 2,4,5-T and the positive TCDD analyses of beef fat. This is not altogether unexpected at the application levels used, *ca.* 1 lb 2,4,5-T/acre with *ca.* one head of cattle per two sprayed acres, assuming there was about 0.1 part per million (ppm) TCDD in the 2,4,5-T. Under these conditions, the accumulation of a few ppt of TCDD in beef fat would correspond to only a small percentage of the amount applied per head. Nevertheless, it is possible that at least some of the TCDD came from now-discontinued industrial operations in Missouri known to have released TCDD into the environment. More analyses of samples from carefully chosen locations may be needed to settle this point.

Meanwhile, we have taken a different and possibly more direct approach to estimating human exposure to TCDD, through the analysis of human milk. This can provide a measure of the level of TCDD intake of the individual. In a preliminary study we analyzed milk samples from 18 women living in areas where 2,4,5-T is used on rangeland or in forestry and 6 women from the Boston area. We found four positive samples (with about 1 ppt each) in the former group and none in the latter. This possible association with the use of 2,4,5-T does not involve a large enough number of samples to be statistically significant. Nevertheless, it has led us, in collaboration with the National Institute of Environmental Health Sciences, to initiate a somewhat larger study, which includes blanks and calibration samples interspersed among the samples from 2,4,5-T areas. Analyses for TCDD in mother's milk on a still larger scale are being undertaken by the EPA using samples from women living near sprayed forests in the Pacific Northwest.

As estimates become available for the level of human exposure to TCDD, more accurate information will be needed regarding the level of chronic exposure which may be toxic. The EPA has attempted to estimate levels below which there is unlikely to be any detrimental effect in man, using laboratory data from long-term feeding of TCDD to rats. This use of long-term exposure data is important because there are indications that the toxic effects of TCDD may be extraordinarily cumulative (7). However, the rat is not a very appropriate species for making extrapolations to man. It is relatively insensitive to the lethal effect of TCDD when compared with other species, such as the guinea pig and, more importantly, the rhesus monkey.

It is already clear from a 9-month feeding experiment that the lethal level for chronic TCDD exposure in monkeys is less than 500 ppt in the diet, possibly much less (8). If TCDD toxicity were completely cumulative in the monkey, the lethal chronic dietary level could be about 20 ppt. Toxicity of a different nature at even lower levels is suggested by a report that TCDD can be carcinogenic to rats at dietary levels as low as 5 ppt (9). Although there is no evidence that anyone in the U.S. is receiving this much TCDD on a steady basis, it is customary to set the permissible level of human exposure to toxic substances very much below the levels found to be lethal or carcinogenic to laboratory animals. Thus, considering the range of uncertainty in both the level of human exposure and the level which might be toxic, it cannot yet be said whether or not current environmental exposure to TCDD poses a serious, widespread hazard. However, progress in analytical methodology and in understanding the toxicology of TCDD is continuing and, if efficiently exploited, should provide a greatly improved perspective on the TCDD problem before much longer.

REFERENCES

1. Baughman, R. W. and Meselson, M. (1973). An improved analysis for 2,3,7,8-tetrachlorodibenzo-*p*-dioxin. In: *Advances in Chemistry Ser. 120* ("Chlorodioxins—Origin and Fate"), E. Blair, Ed., American Chemical Society, Washington, D.C., pp. 92–104.

2. Baughman, R. W. and Meselson, M. (1973). An analytical method for detecting TCDD (dioxin): levels of TCDD in samples from Vietnam. *Environmental Health Perspectives, 5:* 27–35. [DHEW Publication No. (NIH) 74–218.]

3. O'Keefe, P. W., Meselson, M., and Baughman, R. W. (1978). A neutral cleanup procedure for 2,3,7,8-tetrachloro-

dibenzo-*p*-dioxin residues in bovine fat and milk. *Journal of the Association of Official Analytical Chemists.*

4. For a collection of papers on various aspects of the environmental toxicology of TCDD see *Environmental Health Perspectives 5,* 1973. [DHEW Publication No. (NIH) 74–218.]

5. Baughman, R. W. (1974). Tetrachlorodibenzo-*p*-dioxins in the environment. High resolution mass spectrometry at the picogram level. Ph.D. Thesis, Department of Chemistry, Harvard University, Cambridge, Massachusetts.

6. Buser, H. R. (1977). Determination of 2,3,7,8-tetrachlorodibenzo-*p*-dioxin in environmental samples by high-resolution gas chromatography and low resolution mass spectrometry. *Analytical Chemistry 49:* 918–922.

7. Allen, J. R. and Carstens, L. A. (1967). Light and electron microscopic observations in *Macaca mulatta* monkeys fed toxic fat. *Am. J. Vet. Res. 28:* 1513–1526. [The TCDD concentration in the toxic fat used in these experiments was not known at the time. In 1974 we determined it to be 3 ppm by high resolution mass spectrometry. However, this value must be viewed as only approximate due to the possibility of sample heterogeneity.

8. Allen, J. R., Barsotti, D. A., Van Miller, J. P., Abrahamson, L. J., and Lalich, J. J. (1977). Morphological changes in monkeys consuming a diet containing low levels of 2,3,7,8-tetrachlorodibenzo-*p*-dioxin. *Food Cosmet. Toxicol. 15:* 401–410.

9. Van Miller, J. P., Lalich, J. J., and Allen, J. R. (1977). Increased incidence of neoplasms in rats exposed to low levels of 2,3,7,8-tetrachlorodibenzo-*p*-dioxin. *Chemosphere, 6:* 537–544.

Morphological Changes in Monkeys Consuming a Diet Containing Low Levels of 2,3,7,8-tetrachlorodibenzo-*p*-dioxin

J. R. Allen, D. A. Barsotti, J. P. Van Miller, L. J. Abrahamson and J. J. Lalich

Abstract—Female rhesus monkeys fed a diet for 9 months containing 500 parts per trillion (ppt) 2,3,7,8-tetrachlorodibenzo-*p*-dioxin (TCDD) became anemic within 6 months and pancytopenic after 9 months of exposure. The marked thrombocytopenia was associated with widespread hemorrhage. Death occurred in 5 of the 8 animals from the 7th to the 12th month of the experiment at exposure levels of from 2 to 3 μg TCDD/kg body weight. At necropsy, in addition to the extensive hemorrhage there was a decided hypocellularity of the bone marrow and lymph nodes. Hypertrophy, hyperplasia and metaplasia of the epithelium in the bronchial tree, bile ducts, pancreatic ducts, salivary gland ducts and palpebral conjunctivae were observed. Squamous metaplasia and keratinization of the sebaceous glands and hair follicles were present in the skin. Death was attributed to complications from the severe pancytopenia.

RESULTS

Clinical Observations

Following 3 months of TCDD exposure (5.8 ± 0.9 μg total intake per animal) the monkeys developed periorbital

---

Published in *Food and Cosmetics Toxicology*. 1977. 15:401–410. Reprinted by permission.

edema, loss of facial hair and eyelashes, accentuated hair follicles, and dry scaly skin. These changes became more prominent in 6 of the 8 animals by the sixth month of TCDD exposure (11.3 ± 1.7 μg total intake per animal). The majority of the animals showed a decrease in hemoglobin and hematocrit by the sixth month. The hematological changes became more accentuated in 6 of the 7 animals that survived 9 months of TCDD exposure (Table 1). However, blood urea nitrogen, total serum lipid, serum cholesterol, serum glutamic pyruvic transaminase (SGPT), total serum protein and albumin/globulin ratios were not altered appreciably during the experiment except in terminal animals where there was a slight decrease in serum albumin

Table 1. Hemograms of monkeys fed 500 parts per trillion 2,3,7,8-tetrachlorodibenzo-*p*-dioxin (TCDD) for 9 months†

Animal No.	White Blood Cells Initial Terminal ($\times 10^3/cm^3$)		Blood Platelets Initial Terminal ($\times 10^3/cm^3$)		Hemoglobin Initial Terminal (gm%)		Hematocrit Initial Terminal (%)	
7***	11.0	2.3	240	23	13.5	4.0	41.0	12.0
9***	7.9	2.4	360	44	14.1	6.9	43.0	21.5
23	10.3	3.8	380	28	14.6	12.6	45.5	40.0
31	10.6	10.4	400	480	14.1	10.7	45.0	35.5
32*	8.5	4.1	300	50	14.1	6.0	43.0	19.5
38***	9.5	3.8	330	34	13.6	6.6	43.0	22.0
41	7.8	8.0	350	340	13.1	11.5	40.0	44.5
49**	8.6	8.7	260	54	13.7	8.5	43.0	29.5

†Values obtained immediately prior to death or after 12 months on trial

*Died during 28th week of experiment

**Died during 36th week of experiment

***Died during 44th week of experiment

Table 2. Body weights of monkeys fed 500 parts per trillion 2,3,7,8-tetrachlorodibenzo-*p*-dioxin (TCDD) for nine months

Animal No.	Initial body weight (kg)	Terminal body weight (kg)*
7	6.19	5.86
9	6.74	5.74
23	6.62	5.03
31	6.02	5.35
32	5.00	4.22
38	5.63	5.22
41	5.73	5.44
49	5.39	4.46

*Weight at necropsy or at 12 months on trial

and an increase in SGPT. The monkeys also lost weight throughout the experiment even though their food intake was unaltered (Table 2).

One of the animals became severely anemic, thrombocytopenic and leukopenic in the 7th month of exposure (11.2 μg total intake) (Table 1). Prior to death the peripheral blood smears were practically devoid of immature red and white blood cells. During the subsequent four months 4 additional animals experienced similar clinical changes and died. Their peripheral leukocytes and platelets decreased to levels as low as 2,300 and 24,000 per mm^3 from normal values of 7–10,000 and 200–400,000 per mm^3, respectively. The total TCDD exposure of these animals at the time of death was 18.6 ± 1.8 μg per animal. Although all animals were removed from the TCDD diet following 9 months of exposure, the 3 animals that survived for 12 months experienced a continuing loss of hair and periorbital edema. One of the surviving monkeys developed a severe leukopenia and thrombocytopenia by the 12th month (Table 1).

Gross Observations at Necropsy

The animals that died experienced hair loss from all portions of the body. The skin was dry and flaky while the hair follicles about the face became accentuated. Loss of eyelashes and swelling of the upper eyelids along with facial edema and petechiae were present in all animals. There were also large hemorrhagic areas around the nares, on the gums and the surfaces of the buccal cavity. In addition, there were petechial hemorrhages over the entire surface of the body. Subcutaneous edema was particularly prominent in the lower abdominal region and inner surfaces of the thighs. Marked irregularities in the growth of the toenails and fingernails and gangrenous necrosis of the distal phalanges were also recorded.

Ascites was prominent in all of the animals. There was marked distension of the intrahepatic biliary system with edema and dilatation (up to 1 cm in diameter) of the common bile duct. Patency of bile ducts was demonstrated by the flow of bile through the ampulla of Vater and a careful dissection of the intrahepatic ducts. The lymph nodes throughout the abdominal cavity had undergone atrophy. In addition to the pale appearance of all abdominal organs, there were foci of hemorrhage in the adrenals, pancreas, liver, endometrium, serosal, and mucosal surface of the gastrointestinal tract and in the urinary bladder.

In the thoracic cavity, the lungs exhibited focal hemorrhage in all lobes. There was bilateral ventricular dilatation, cardiac enlargement and edema. Focal areas of hemorrhage were present in the epicardium, the myocardium, and the endocardium. The skeletal musculature, particularly in the extremities, was edematous, pale and discolored by hemorrhagic foci. Extensive hemorrhage was found in the meninges, as well as isolated hemorrhagic areas in the brain parenchyma. The bone marrow had the appearance of fatty tissue in which there were focal areas of hemorrhage.

Microscopic Observations

The major microscopic changes may be separated into three categories: degeneration or atrophy of the bone marrow and lymphopoietic tissues, hemorrhage, and those related to cellular hypertrophy, hyperplasia and metaplasia.

The bone marrow displayed a decided hypocellularity. The hematopoietic cells of the marrow were replaced primarily by fat cells with focal areas in which lymphoid appearing cells predominated. There was also a conspicuous reduction of discernible erythroid or myeloid stem cells as well as megakaryocytes in the marrow. The lymph nodes throughout the body were hypocellular. In addition to the absence of any distinct lymphoid germinal centers, the cortical lymphocytes were sparse while the medullary cords were narrow or inapparent. Cellular debris and large vacuolated cells predominated in the sinuses. The spleen also was devoid of any distinct lymphoid germinal centers and the small lymphocytes were widely dispersed.

The second major category of microscopic changes was related to the hemorrhage that occurred in tissues throughout the body. Well-circumscribed focal areas of hemorrhage in the heart, lungs, liver, adrenal, pancreas, and skeletal muscle consisted of collections of intact red blood cells that partially disrupted the architectural pattern of the affected tissue. The acute nature of the hemorrhage was manifested by the lack of tissue necrosis and an absence of reactive cells. Petechial areas were prominent in the dermis of the skin, submucosa of the urinary bladder, and in the epithelium of the alimentary tract. Hemorrhages of the endometrium and meninges were more diffuse in their distribution. Hemorrhages in the brain were limited to the Verchow-Robbin's spaces surrounding the blood vessels. The hemorrhage in the bone marrow varied considerably, being diffuse in some areas and focal in others.

Considerable modification in the morphological features of epithelium occurred in the TCDD treated animals. Metaplastic changes characterized by numerous mucous secreting cells were present in the ductal epithelium of the salivary glands, bile ducts, and pancreatic ducts. Similar changes also occurred in the bronchial epithelium and in the palpebral conjunctivae. In addition to the metaplasia, the epithelium of the bile ducts and palpebral conjunctivae developed considerable hypertrophy and hyperplasia. Squamous metaplasia of the sebaceous glands occurred more extensively in the skin of the face and to a lesser extent in other areas.

Hypertrophy, hyperplasia, and metaplasia were observed in the gastric epithelium where the parietal and zymogenic cells were replaced by those that secreted mucous. In addition to the thickened mucosal lining, many of the epithelial cells invaded the submucosa through interruptions in the muscularis mucosa. Such ectopic epithelium assumed many different patterns. Some isolated cells encompassed large mucinous cysts, while others were arranged in sheets or formed acinar structures. Considerable edematous fluid usually surrounded the epithelium. Inflammatory cells with a predominance of neutrophils were seen in close proximity to ruptured submucosal mucinous cysts. Rupture of such cysts produced the ulcers that were found in the gastric mucosa. Hypertrophy and hyperplasia were also apparent in the transitional epithelium throughout the urinary tract.

In addition to squamous metaplasia in the epithelium of the sebaceous glands, there was hyperkeratosis of the skin with keratinization of the hair follicles and the adjacent sebaceous glands. The latter alterations were particularly prominent in the Meibomian glands of the eyelids. The pronounced thickening of the fingernails and toenails is considered to be secondary to excessive keratin production.

In the heart, petechial hemorrhages were prominent in the atria and in the tips of the papillary muscles. Additionally, foci of subendothelial and pericardial hemorrhage were frequently found in the ventricles. In two of the five hearts the erythrocytes were hemolyzed so that heme pigments and hemosiderin could be identified by Perl's stain in the intracellular spaces. There was intra- and intercellular edema as characterized by separation of the muscle fibers and myofibrils.

DISCUSSION

The data presented in this report indicate that profound cellular alterations are induced in many tissues following the ingestion of minute concentrations of TCDD by primates for 9 months. Perhaps of foremost importance is the decided effect of TCDD on the hematopoietic system. As exposure time lengthens, the severity of the cellular deterioration in bone marrow and lymphoid tissue becomes widespread. Terminally the animals developed severe pancytopenia. The hematological data presented here agree quite well with the previous studies conducted in this laboratory where mixtures of dioxins, including TCDD, were fed to nonhuman primates (Allen & Carstens, 1967). In the previous investigation the monkeys also developed anemia and leukopenia prior to death. Similar hematological changes have been recorded in rats given a single dose of 50 μg of TCDD per kg body weight (Van Miller & Allen, 1977). In addition to the depletion of circulating lymphocytes, atrophy of the thymus, spleen, and lymph nodes has been observed in several species. These changes may be related to the immune suppression which has been recorded in TCDD exposed animals. Sublethal doses of TCDD have been shown to suppress cell mediated and humoral immunity in guinea pigs (Vos, Moore & Zinkl, 1973).

There are other complications that arise as a result of the decrease in hematopoiesis. The widespread hemorrhage that was observed in the monkeys prior to death corresponded to the decrease in blood platelets. It is logical to assume that the hemorrhage at least in part was associated with the altered clotting capability of the blood. The ventricular dilatation and myocardial hypertrophy in the TCDD exposed monkeys may be related in part to the reduction of circulating erythrocytes and subsequent increase in the cardiac workload.

In this experiment it was shown that ingestion of 2 to 3 μg per kg body weight of TCDD over a 9 month period was capable of producing either cellular destruction or alteration in organs sufficiently great to cause death in over 50% of the nonhuman primates. It is of interest that the LD-50 of monkeys exposed to a single oral dose of TCDD is 50–70 μg per kg body weight (McConnell, Moore & Dalgard, 1977) with death occurring within 45 days. There are some major differences in the changes that occur in monkeys acutely and chronically exposed to TCDD. While death was associated with severe pancytopenia in the chronically exposed animals, hematological changes were not a constant feature of those acutely exposed. In addition, widespread hemorrhage occurred only in the chronically exposed animals. Changes after acute or chronic exposure which were similar are: loss of hair and nails, keratinization of the Meibomian glands and hair follicles, hyperplasia of the transitional epithelium in the renal pelvis, mucous epithelium of the stomach and bile duct, a decrease in serum albumin, and an increased SGPT.

The morphological changes that have been recorded in monkeys exposed to TCDD indicate that it is capable of suppressing hematopoiesis and inducing cellular alterations in epidermal appendages and mucosal epithelium of the biliary, pancreatic, intestinal, and urinary tract. After exposure to TCDD the cell population in the bone marrow and lymphoid elements underwent degeneration and necrosis. Other less specialized cells such as the bile duct epithelium and mucous cells of the stomach were not injured as severely. However, as a result of injury the latter cells made attempts to compensate by undergoing hypertrophy and/or hyperplasia. Some of the injured cells also reverted to more primitive forms that were capable of surviving in a less desirable environment. This was seen in the sebaceous glands where the secretory and ductal epithelium reverted to squamous cells.

In light of the observations made on the monkeys employed in this experiment, there are areas that warrant special attention in animals including man exposed to TCDD. Perhaps of foremost importance are the alterations that occur in the hematopoietic tissue. Anemia, thrombocytopenia, and leukopenia were the most debilitating changes in the primates. The altered lymphopoiesis could be associated with immune suppression. The possibility of reproductive abnormalities also exists. Altered levels of serum progesterone and estradiol associated with difficulties in conception and early abortions have been observed in female monkeys exposed to low levels of TCDD (Allen, Van Miller, Barsotti & Abrahamson, 1977). Testicular atrophy has also developed in male monkeys given small amounts of dioxins (Allen & Carstens, 1967). In addition, the widespread hypertrophy, hyperplasia and metaplasia that occurred in the epithelium of monkeys exposed to TCDD, and a greater frequency of tumors in rats, suggests a carcinogenic action of TCDD.

REFERENCES

Allen, J. R. & Carstens, L. A. (1967). Light and electron microscopic observations in *Macaca mulatta* monkeys fed toxic fat. *Am. J. Vet. Res. 28,* 1513.

Allen, J. R., Van Miller, J. P., Barsotti, D. A., & Abrahamson, L. J. (1977). Morbidity and mortality in primates resulting from picogram doses of 2,3,7,8-tetrachlorodibenzo-*p*-dioxin. *Nature,* submitted.

Carter, C. D., Kimbrough, R. D., Liddle, J. A., Cline, R. E., Zack, M. M., & Barthel, W. F. (1975). Tetrachlorodibenzodioxin: an accidental poisoning episode in horse arenas. *Science 188,* 738.

McConnell, E. E., Moore, J. A., & Dalgard, D. W. (1977). Toxicity of 2,3,7,8-tetrachlorodibenzo-*p*-dioxin (TCDD) in rhesus monkeys: a single oral dose. *Toxicol. Appl. Pharmacol.,* submitted.

Rose, H. R. & Rose, S. P. R. (1972). Chemical spraying as reported by refugees from South Vietnam. *Science 177,* 710.

Van Miller, J. P. & Allen, J. R. (1977). Chronic toxicity of 2,3,7,8-tetrachlorodibenzo-*p*-dioxin in rats. *Fed. Proc. 36,* 396.

Vos, J. G., Moore, J. A. & Zinkl, J. (1973). Effect of TCDD on the immune system of laboratory animals. *Environ. Health Persp. 4,* 149.

Increased Incidence of Neoplasms in Rats Exposed to Low Levels of 2,3,7,8-tetrachlorodibenzo-*p*-dioxin

J. P. Van Miller, J. J. Lalich and J. R. Allen

INTRODUCTION

2,3,7,8-tetrachlorodibenzo-*p*-dioxin (TCDD) and other chlorinated dioxins have been shown to cause a wide variety of cellular alterations in experimental animals. Lesions including hydropericardium and ascites in chickens[1,2], as well as liver necrosis and atrophy of the lymphoid organs in rats[3,4], mice and guinea pigs[5] have been observed following the administration of doses near LD_{50} values. Similar effects have been observed in rats given TCDD on a daily basis for 13 weeks[6]. Acne, alopecia, liver necrosis and anemia have been recorded in nonhuman primates exposed to a variety of chlorinated dioxins[7]. More recently, rhesus monkeys fed 500 parts per trillion TCDD in their diets on a daily basis developed signs of toxicity at intake levels as low as 1 μg/kg[8]. A total intake of 3 μg/kg over 9 months was sufficient to cause death in 5 of 8 animals fed the experimental diet. Lesions observed included a severe pancytopenia due to suppression of bone marrow function as well as metaplasia of ductal epithelium throughout the body.

Chemosphere Vol. 6, No. 9, pp. 537–544, 1977. Pergamon Press. Printed in Great Britain. Reprinted by permission.

Accidental exposure of man to TCDD has resulted in chloracne, cystitis, pyelonephritis, fatigue and peripheral numbness[9,10,11]. In addition, an increased incidence of liver cancer attributed to TCDD has been reported in Vietnamese exposed to the defoliant Agent Orange[12].

In the present study rats were fed various levels of TCDD in their diets to determine the toxicity and possible carcinogenicity from low level exposure to this compound.

MATERIALS AND METHODS

Male Sprague-Dawley rats (Sprague-Dawley, Madison, Wisconsin U.S.A.) weighing approximately 60 g were divided into 10 equal groups (10 animals per group). The animals were housed 2 to a cage and fed ground chow (Purina Rat Chow, Ralston-Purina Co., St. Louis, Missouri U.S.A.) for two weeks. After this period of acclimation, the animals were placed on a diet containing one of the following concentrations of TCDD: 1, 5, 50, 500 parts per trillion (ppt; 10^{-12} g TCDD/g food); 1, 5, 50, 500 or 1000 parts per billion (ppb; 10^{-9} g TCDD/g food). In addition, one group was maintained on the pre-experimental diet to serve as controls.

The experimental diets were prepared in 5 kg batches as follows. The appropriate quantity of TCDD was suspended in 0.5 ml of acetone and dissolved in 100 ml of corn oil. The corn oil was added to approximately 1 kg of ground chow and mixed thoroughly. This premix was then mixed with the remaining chow to yield 5 kg. The diets were stored at —5 °C to prevent spoilage.

Two animals from each group were isolated and their food intake monitored for 2 weeks. Weights were recorded for all of the animals three times weekly for the first month and every two weeks thereafter. The animals were not manipulated in any other way until the 65th week of the experiment when laparotomies were performed on all of the surviving animals. At the time of laparotomy biopsies were obtained from any tumors observed. The animals were maintained on the diets for 78 weeks. At that time the experimental animals were placed on the control diet. At 95 weeks all surviving animals were sacrificed.

At death or sacrifice, complete necropsies were performed, and samples of tissues were taken for microscopic examination. Tissues were fixed in 10% neutral buffered formalin, embedded in paraffin, cut in 5 micron sections, stained with hematoxylin and eosin and examined by light microscopy. Biopsy samples were treated similarly. Additional staining methods including mucicarmine, Gomori reticulum stain, PTAH, and Masson's trichrome were used to aid in the diagnosis of neoplasms[13].

RESULTS

The three highest dose levels (50, 500 and 1000 ppb) produced acute toxicity in all of the animals. Food intake in these groups (10 ± 4 g/day) was significantly lower than for controls (21 ± 2 g/day). None of the animals in these three groups gained weight after the start of the experimental diet. All of the animals died between the second and fourth week of the experiment (Table 1).

At necropsy, atrophy of the thymus and spleen, dilatation of the common bile ducts and hemorrhage in the gastrointestinal tract were observed. Microscopically, there was severe liver necrosis as well as cellular proliferation in the bile ducts. There was cellular hyperplasia of the mucosa and edema in the wall of the common bile ducts. In addition, decreased spermatogenesis was observed in many of the animals.

Table 1. Mortality in Rats Fed Various Levels of 2,3,7,8-tetra-chlorodibenzo-*p*-dioxin on a Daily Basis

Level of TCDD in the Diet	Approximate Weekly Dose per animal ($\mu g/kg$ body wt)	No. Animals Dead at 95 Weeks[a]	Week of First Death
0	—	6	68
1 ppt[b]	0.0003	2	86
5 ppt	0.001	4	33
50 ppt	0.01	4	69
500 ppt	0.1	5	17
1 ppb[c]	0.4	10	31
5 ppb	2.0	10	31
50 ppb	24	10	3
500 ppb	240	10	2
1000 ppb	500	10	2

[a] 10 animals per group; surviving animals sacrificed at 95 weeks
[b] 1 ppt = 1 part per trillion = 10^{-12} g TCDD/g food
[c] 1 ppb = 1 part per billion = 10^{-9} g TCDD/g food

In contrast to those animals receiving the three highest dose levels, the food intake for those on the other dose levels was similar to the controls (20±2 g/day). Weight gain was significantly different from that of the controls only in the 5 ppb group. The maximum weight of the animals in this group was 391±54 as compared to 531±44 in the control group. Only one animal (500 ppt) died before the 30th week of the experiment. All of the animals in the 5 and 1 ppb groups died by the 90th week of the experiment. The mortality figures in the control, 1 ppt, 5 ppt, 50 ppt and 500 ppt groups were 60, 20, 40, 40, and 50%, respectively, after 95 weeks (Table 1).

All of the control and 1 ppt animals, including those that died prior to sacrifice, had extensive senile degenerative changes of the kidney. Similar lesions were observed in the sacrificed animals from the other 5 groups as well as in 10 of the animals that died from these groups. Three of the animals in the 5 ppb group died of aplastic anemia. One of the animals in the 500 ppt group had severe liver infarction.

Nineteen (57%) of the animals that died in the 6 groups fed subacute levels of TCDD had neoplastic alterations. Biopsies taken at the 65 week laparotomies revealed 2 additional neoplasms (5 ppt, 5 ppb). Four of the sacrificed animals also had neoplastic alterations. The overall incidence of neoplasms in the 6 experimental groups was 38% with no neoplasms observed in the 1 ppt group. Five of the 23 animals with tumors had two primary neoplastic lesions (Table 2).

The squamous cell lung tumors observed in the animals from the 5 ppb group were characterized by squamous metaplasia of the bronchial epithelium and massive keratinization. The ear duct carcinoma (5 ppt) had similar metaplasia of ductal epithelium and keratinization. The lymphocytic leukemia was associated with a massive enlargement of the spleen (12.5 g). The most malignant neoplasms observed were the 3 retroperitoneal histiocytomas (5 ppt, 1 ppb) from which hematogenous metastases were observed to the lungs, kidneys, liver and skeletal musculature. The angiosarcoma in the animal from the 5 ppt group and one of the cholangiocarcinomas in an animal from the 5 ppb group were diagnosed from biopsies taken at the 65 week laparotomies. No neoplasms were observed in any of the control animals. In addition, 40 male Sprague-Dawley rats used as controls for a separate experiment that were received at the same time and housed under identical conditions showed no neoplastic changes when sacrificed at 18 months.

Table 2. Summary of Neoplastic Alterations Observed in Rats Fed Subacute Levels of 2,3,7,8-tetrachlorodibenzo-*p*-dioxin for 78 Weeks

Level of TCDD	No. of Animals with Neoplasms[a]	No. of Neoplasms	Diagnosis
0	0	0	—
1 ppt[b]	0	0	—
5 ppt	5	6	1 ear duct carcinoma 1 lymphocytic leukemia 1 adenocarcinoma (kidney) 1 malignant histiocytoma (peritoneal)[d] 1 angiosarcoma (skin) 1 Leydig cell adenoma (testes)
50 ppt	3	3	1 fibrosarcoma (muscle) 1 squamous cell tumor (skin) 1 astrocytoma (brain)
500 ppt	4	4	1 fibroma (striated muscle) 1 carcinoma (skin) 1 adenocarcinoma (kidney) 1 sclerosing seminoma (testes)
1 ppb[c]	4	5	1 cholangiocarcinoma (liver) 1 angiosarcoma (skin) 1 glioblastoma (brain) 2 malignant histiocytomas (peritoneal)[d]
5 ppb	7	10	4 squamous cell tumors (lung) 4 neoplastic nodules (liver) 2 cholangiocarcinomas (liver)

[a]10 animals per group
[b]1 ppt = 1 part per trillion = 10^{-12} g TCDD/g food
[c]1 ppb = 1 part per billion = 10^{-9} g TCDD/g food
[d]Metastases observed

DISCUSSION

The administration of TCDD to rats on a daily basis at levels of 50 ppb and above induced signs of toxicity similar to those previously reported[1-8]. The two week period required for TCDD to cause mortality is consistent with previous reports[4]. The cause for this latent period remains an unanswered but potentially important question in the determination of the mode of action of TCDD.

The high incidence of neoplasms in rats fed subacute levels of TCDD suggests the carcinogenic potential of the compound. The tumors of ductal epithelium (e.g., lung, sebaceous gland) and the cholangiocarcinomas are consistent with findings in nonhuman primates fed TCDD where metaplasia of the ductal epithelium in the lung and sebaceous glands of the skin as well as proliferation of the epithelium of the bile ducts were observed[8]. Additionally, the ear duct tumor (5 ppt) is similar in location and morphology to that previously described in the auditory sebaceous gland (Zymbal's gland) in rats fed the carcinogen 2-acetylaminofluorene[14].

The possibility that TCDD is a potent promoter of neoplastic changes rather than an inducer cannot be overlooked. Neoplasms in the liver, the primary site of TCDD localization[15], occurred only in animals fed 1 and 5 ppb TCDD in their diets. The wide variety of neoplasms observed in this study is not consistent with many of the known chemical carcinogens. In addition, reports of increased liver cancer in Vietnamese 8 to 10 years following exposure to defoliants containing TCDD[12] tends to support a promotion mechanism for neoplasms caused by TCDD.

While this study does not prove conclusively that TCDD is a carcinogen, the increased incidence of neoplastic alterations in rats fed levels of TCDD as low as 5 ppt is of obvious concern. The elucidation of the molecular mechanisms of TCDD toxicity is required for a better understanding of these results.

ACKNOWLEDGMENTS

This investigation was supported in part by U.S. Public Health Service grants ES00472 and RR00167 and the University of Wisconsin Sea Grant Program. Part of this research was conducted in the University of Wisconsin-Madison Biotron, a controlled environmental research facility supported by the National Science Foundation and the University of Wisconsin. Primate Center Publication No. 17-002.

REFERENCES

1. D. Firestone, Environ. Health Persp. *5,* 59 (1973).

2. J. R. Allen, Amer. J. Vet. Res. *25,* 1210 (1964).

3. B. A. Schwetz, J. M. Norris, G. L. Sparschu, V. K. Rowe, P. J. Gehring, J. L. Emerson, and C. G. Gerbig, Environ. Health Persp. *5,* 87 (1973).

4. J. R. Allen, J. P. Van Miller, and D. H. Norback, Food Cosmet. Toxicol. *13,* 501 (1975).

5. B. N. Gupta, J. G. Vos, J. A. Moore, J. G. Zinkl, and B. C. Bullock, Environ. Health Persp. *5,* 125 (1973).

6. R. J. Kociba, P. A. Keeler, C. N. Park, and P. J. Gehring, Toxicol. Appl. Pharmacol, *35,* 553 (1976).

7. J. R. Allen and L. A. Carstens, Amer. J. Vet. Res. *38,* 1513 (1967).

8. J. R. Allen, D. A. Barsotti, J. P. Van Miller, L. J. Abrahamson, and J. J. Lalich, Food Cosmet. Toxicol., in press (1977).

9. G. May, Brit. J. Ind. Med. *30,* 276 (1973).

10. C. D. Carter, R. D. Kimbrough, J. A. Liddle, R. E. Cline, M. M. Zack, W. F. Barthel, Science *188,* 738 (1975).

11. H. R. Rose and S. P. R. Rose, Science *177*, 710 (1972).
12. T. T. Tung, Chirurgie *99*, 427 (1973).
13. Armed Forces Institute of Pathology, Manual of Histologic and Special Staining Techniques. McGraw-Hill, New York (1960).
14. J. D. Laws, G. Rudali, R. Royer, and P. Mabille, Cancer Res. *15*, 139 (1955).
15. J. P. Van Miller, R. J. Marlar, and J. R. Allen, Food Cosmet. Toxicol. *14*, 31 (1976).

Bibliography

Allen, J. R.; Abrahamson, L. J.; and Norback, D. H. 1973. Biological effects of polychlorinated biphenyls and triphenyls on the subhuman primate. *Environ. Res.* 6:344–54.

Allen, J. R.; Barsotti, D. A.; Lambrecht, L. K.; and Van Miller, J. P. 1978. Reproductive effects of halogenated aromatic hydrocarbons on non-human primates. Paper read at Intern. Conf. on Health Effects of Halogenated Aromatic Hydrocarbons. 21–30 June at New York Academy of Sciences.

Allen, J. R.; Barsotti, D. A.; and Van Miller, J. P. 1977. Reproductive dysfunction on nonhuman primates exposed to dioxins. *Toxicol. Appl. Pharmacol.* 41:177.

Allen, J. R.; Barsotti, D. A.; Van Miller, J. P.; Abrahamson, L. J.; and Lalich, J. J. 1977. Morphological changes in monkeys consuming a diet containing low levels of 2,3,7,8-tetrachlorodibenzo-*p*-dioxin. *Food Cosmet. Toxicol.* 15:401-10.

Allen, J. R., and Carstens, L. A. 1966. Electron microscopic alterations in the liver of chickens fed toxic fat. *Lab. Invest.* 15(6):970–79.

Allen, J. R.; Carstens, L. A.; and Barsotti, D. A. 1974. Residual effects of short-term, low-level exposure of nonhuman primates to polychlorinated biphenyls. *Toxicol. Appl. Pharmacol.* 30:440–51.

Allen, J. R.; Van Miller, J. P.; and Norback, D. H. 1975. Tissue distribution, excretion and biological effects of [^{14}C] tetrachlorodibenzo-*p*-dioxin in rats. *Food Cosmet. Toxicol.* 13:501–15.

Armstrong, R. W.; Eichner, E. R.; Klein, D. E. et al. 1969. Pentachlorophenol poisoning in a nursery for newborn infants. *J. Pediatrics* 75:317–25.

Arsenault, R. D. 1976. Pentachlorophenol and contained chlorinated dibenzodioxins in the environment; a study of environmental fate, stability, and significance when used in wood preservation. Am. Wood Preservers' Assoc.

Axelson, O., and Sundell, L. 1974. Herbicide exposure, mortality and tumor incidence. An epidemiological investigation on Swedish railroad workers. *Work Environ. Health* 11:21–28.

Bage, G.; Cekanova, E.; and Larson, K. S. 1973. Teratogenic and embryotoxic effects of the herbicides di- and trichlorophenoxyacetic acids (2,4-D and 2,4,5-T). *Acta pharmacol. et toxicol.* 32:408–16.

Bauer, H.; Schulz, K. H.; and Spiegelberg, U. 1961. Occupational intoxication in the production of chlorinated phenol compounds. Translated from German. *Arch. Indust. Pathol. Indust. Hyg.* 18:538–55.

Baughman, R. W. 1974. Tetrachlorodibenzo-*p*-dioxins in the environment. High resolution mass spectrometry at the picogram level. Ph.D. dissertation, Harvard Univ.

Baughman. R. W. and Meselson, M. S. 1973. An analytical method for detecting TCDD (dioxin): levels of TCDD in samples from Vietnam. *Environ. Health Persp.* 5:27–34.

Beatty, P. W.; Lembach, K. J.; Holscher, M. A.; and Neal, R. A. 1975. Effects of 2,3,7,8-tetrachlorodibenzo-*p*-dioxin (TCDD) on mammalian cells in tissue cultures. *Toxicol. Appl. Pharmacol.* 31:309–12.

Bianotti, F. 1977. Chloracne from tetrachloro-2,3,7,8-dibenzo-*p*-dioxin in children. Translated from *Ann. Dermatol. Venereol.* (Paris). 104:825–29.

Bowes, G. W.; Simoneit, B. R.; Burlingame, A. L.; de Lappe, B. W.; and Risebrough, R. W. 1973. The search for chlorinated dibenzofurans and chlorinated dibenzodioxins in wildlife population showing elevated levels of embryonic death. *Environ. Health Persp.* 5:191–98.

Buser, H. R.; Bosshardt, H.-P.; and Rappe, C. 1978. Identification of polychlorinated dibenzo-*p*-dioxin isomers found in fly ash. *Chemosphere* 7:165–72.

Buser, H. R.; Bosshardt, H.-P.; Rappe, C.; and Lindahl, R.

1978. Identification of polychlorinated dibenzofuran isomers in fly ash and PCB pyrolyses. *Chemosphere*. In press.

Buu-Hoi, N. P.; Chanh, Pham-Huu; Sesque, G.; Azum-Gelade, M. C.; and Saint-Ruf, G. 1972. Organs as targets of "dioxin" (2,3,7,8-tetrachlorodibenzo-*p*-dioxin) intoxication. *Sonderdruck aus Die Naturwissenschaften* 4:174–75.

Buu-Hoi, N. P.; Saint-Ruf, G.; Bigot, P.; and Mangane, M. 1971. Preparation, propriétés et identification de la "dioxine" (tetrachloro-2,3,7,8 dibenzo-*p*-dioxin) dans les pyrolysats de defoliants à base d'acide trichloro-2,4,5 phenoxyacetique et de ses esters et des vegetaux contaminés. Academy of Sciences (Paris). Ser. D. 273:708–11.

Carter, C. D.; Kimbrough, R. D.; Liddle, J. A.; Cline, R. E.; Zack, M. M., Jr.; Barthel, W. F.; Koehler, R. E.; and Phillips, P. E. 1975. Tetrachlorodibenzodioxin: an accidental poisoning episode in horse arenas. *Science* 188:738–40.

Cattabeni, F.; Cavallaro, A.; and Galli, G., eds. 1978. *Dioxin: toxicological and chemical aspects*. SP Medical and Scientific Books. New York: John Wiley, Halsted Press.

Clark, D. E.; Palmer, J. S.; Radeleff, R. D.; Crookshank, H. R.; and Farr, F. M. 1975. Residues of chlorophenoxy acid herbicides and their phenolic metabolites in tissues of sheep and cattle. *J. Agr. Food Chem.* 23(3):571–78.

Commoner, B., and Scott, R. E. November 1976. *U.S. Air Force Studies on the Stability and Ecological Effects of TCDD (Dioxin): An Evaluation Relative to the Accidental Dissemination of TCDD at Seveso, Italy*. Center for the Biology of Natural Systems, Washington Univ., St. Louis.

——. 1976. Accidental contamination of soil with dioxin in Missouri: effects and countermeasures. Center for the Biology of Natural Systems. St. Louis.

Courtney, K. D. 1976. Mouse teratology studies with chlorodibenzo-*p*-dioxins. *Bull. Environ. Contam. Toxicol.* 16(6): 674–81.

Courtney, K. D., and Moore, J. A. 1971. Teratology studies with 2,4,5-trichlorophenoxyacetic acid and 2,3,7,8-tetrachlorodibenzo-*p*-dioxin. *Toxicol. Appl. Pharmacol.* 20:396–403.

Crosby, D. G., and Wong, A. S. 1976. Photochemical generation of chlorinated dioxins. *Chemosphere* 5:327–32.

———. 1977. Environmental degradation of 2,3,7,8-TCDD. *Science* 195:1337–38.

Crow, K. D. 9 July 1977. Effects of dioxin exposure (letter to the editor). *The Lancet* 2:82.

Crummett, W. B., and Stehl, R. H. August 1973. The determination of chlorinated dibenzo-*p*-dioxins and dibenzofurans in various materials. *Environ. Health Persp.* Experimental Issue No. 5.

Cunningham, H. M., and Williams, D. T. 1972. Effect of tetrachlorodibenzo-*p*-dioxin on growth rate and the synthesis of lipids and proteins in rats. *Bull. Environ. Contam. Toxicol.* 7:45–51.

DiGiovanni, J. et al. 1977. Tumor-initiating ability of 2,3,7,8-tetrachlorodibenzo-*p*-dioxin (TCDD) and arochlor 1254 in the two-stage system of mouse skin carcinogenesis. *Bull. Environ. Contam. Toxicol.* 18:552–56.

Dioxin Symposium Abstracts. 162nd Am. Chem. Soc. Meeting, 12–17 September 1971 at Washington, D.C.

Dow Chemical Co. 1974. *Industrial vegetation management.* 6(1).

Dow Chemical Co. July 1975. *Chemicals, human health and the environment: a collection of Dow scientific papers.* Vol. 1.

Epstein, S. S. 1978. *The Politics of Cancer.* Sierra Club Books.

Faith, R. E., and Moore, J. A. 1977. Impairment of thymus-dependent immune functions by exposure to the developing immune system to 2,3,7,8-tetrachlorodibenzo-*p*-dioxin (TCDD). *J. Toxicol Environ. Health* 3:451–64.

Fara, G. M. (Coordinator). 27 August 1977. Medical-Epidemiological Commission, Regione Lombardia, Giunta Regionale, Milan.

Firestone, D. 1977. The TCDD problem: a review. In *Symposium on chlorinated phenoxy acids and their dioxins; mode of action, health risks and environmental effects.* Stockholm: Royal Swedish Academy of Sciences.

Fujita, K.; Fujita, H.; and Funazaki, A. 1975. Chromospheric abnormalities brought about by the use of 2,4,5-T. Translated from Japanese. *J. Jap. Assoc. Rural Med.* 24(2):77–79.

Galston, A. W. 1971. Some implications of the widespread use of herbicides. *Bioscience* 21:891–92.

———. 25 November 1967. Herbicides in Vietnam. *New Republic.* 19–21.

———. 1972. Science and social responsibility: a case history. *Ann. N.Y. Acad. Sci.* 196:223–35.

Goldstein, J. A.; Hickman, P.; Bergman, H.; and Vos, J. G. 1973. Hepatic porphyria induced by 2,3,7,8-tetrachlorodibenzo-*p*-dioxin. *Res. Commun. Chem. Pathol. Pharmacol.* 6(3):919–28.

Greig, J. B.; Jones, G.; Butler, W. H.; and Barnes, J. M. 1973. Toxic effects of 2,3,7,8-tetrachlorodibenzo-*p*-dioxin. *Food Cosmet. Toxicol.* 11:585–95.

Greig, J. B.; Taylor, D. M.; and Jones, J. D. 1974. Effects of 2,3,7,8-tetrachlorodibenzo-*p*-dioxin on stimulated DNA synthesis in the liver and kidney of the rat. *Chem.-Biol. Interactions* 8:31–39.

Gupta, B. N.; Vos, J. G.; Moore, J. A.; Zinkl, J. G.; and Bullock, B. C. August 1973. Pathologic effects of 2,3,7,8-tetrachlorodibenzo-*p*-dioxin in laboratory animals. *Environ. Health Persp.* Experimental Issue No. 5.

Halling, H. 1978. Suspected link between exposure to hexachlorophene and malformed infants. Paper read at Intern. Conf. on Health Effects of Halogenated Aromatic Hydrocarbons. 21–30 June at New York Academy of Sciences.

Harris, M. W.; Moore, J. A.; Vos, J. G.; and Gupta, B. N. 1973. General biologic effects of TCDD in laboratory animals. *Environ. Health Persp.* 5:101–09.

Hay, A. 1976. Toxic cloud over Seveso. *Nature* 262:636–38.

———. 1976. Seveso: the aftermath. *Nature* 263:538–40.

———. 1977. Seveso: dioxin damage. *Nature* 266:7–8.

———. 1977. Identifying carcinogens. *Nature* 269:468–70.

———. 1977. Tetrachlorodibenzo-*p*-dioxin release at Seveso. *Disasters* 1(4):289–308. London: Pergamon Press.

Huff, J. E., and Wassom, J. S. August 1973. Chlorinated dibenzodioxins and dibenzofurans—an annotated literature collection. *Environ. Health Persp.* Experimental Issue No. 5.

Hussain, S.; Ehrenberg, L.; Lofroth, G.; and Gejvall, T. 1972. Mutagenic effects of TCDD on bacterial systems. *Ambio* 1(1):32–33.

Jensen, N. E., and Walker, A. E. 1972. Chloracne: three cases. *Proc. Roy. Soc. Med.* 65:687–88.

Jensen, S., and Renberg, L. April 1972. Contaminants in pentachlorophenol: chlorinated dioxins and predioxins (chlorinated hydroxy-diphenylethers). *Ambio* 1(2):62–65.

Jirasek, L.; Kalensky, J.; Kubec, K.; Pazderova, J.; and Lukas, E. 1974. Acne chlorina, poryphyria cutanea tarda a jine projevy celkove intoxikace pri vyrobe herbicid. *Cs. Derm.* 49:145–57.

Jirasek, L. et al. 1976. Chlorakne-porphyria cutanea tarda und andere Intoxikationen durch herbizide. *Hautarzt* 27:328.

Johnson, J. E. 1971. The public health implications of widespread use of the phenoxy herbicides. *Bioscience* 21:899–905.

Khera, K. S., and Ruddick, J. A. 1973. Polychlorodibenzo-*p*-dioxins: perinatal effects and the dominant lethal test in Wistar rats. In *Chlorodioxins—origin and fate.* Ed., E. A. Blair. Advances in Chemistry Ser., No. 120, pp. 70–84. Washington, D.C.: Am. Chem. Soc.

Kimbrough, R. D., and Gaines, T. B. 1971. Hexachlorophene effects on the rat brain. *Arch. Environ. Health* 23:114–18.

Kimbrough, R. D. 1972. Toxicity of chlorinated hydrocarbons and related compounds: a review including chlorinated dibenzodioxins and chlorinated dibenzofurans. *Arch. Environ. Health* 25:125–31.

Kimbrough, R. D. 1974. The toxicity of polychlorinated polycyclic compounds and related chemicals. *CRC Crit. Rev. Toxicol.* 2:445–99.

Kimbrough, R. D. 1976. Pharmacodynamics and neurotoxicity of hexachlorophene including ultrastructure of the brain lesion. *Clinic. Toxicol.* 9(6):969–79.

Kimbrough, R. D. 1978. The carcinogenic and other chronic

effects of persistent halogenated organic compounds. Paper read at Intern Conf. on Health Effects of Halogenated Aromatic Hydrocarbons, 21–30 June at New York Academy of Sciences.

Kimbrough, R. D.; Carter, C. D.; Liddle, J. A.; Cline, R. E.; and Phillips, P. E. 1977. Epidemiology and pathology of a tetrachlorodibenzodioxin poisoning episode. *Arch. Environ. Health* 32:77–86.

Kimmig, J., and Schulz, K. H. 1957. Chlorinated aromatic cyclic ethers as the cause of chloracne. *Naturwiss.* 44:337–38.

Kleu, G., and Goltz, R. 1971. Late and long-term injuries following the chronic occupational action of chlorophenol compounds: catasmnetic neurologic-psychiatric and psychological studies. Translated from German. *Med. Klin.* (Munich) 66(2):53–58.

Kociba, R. J. 1976. 2,3,7,8-tetrachlorodibenzo-*p*-dioxin (TCDD): results of a 13-week oral toxicity study in rats. *Toxicol. Appl. Pharmacol.* 35:553–74.

Kociba, R. J.; Keeler, P. A.; Park, C. N.; and Gehring, P. J. 1976. 2,3,7,8-tetrachlorodibenzo-*p*-dioxin (TCDD): results of a 13-week oral toxicity study in rats. *Toxicol. Appl. Pharmacol.* 35:553–74.

Kociba, R. J.; Keyes, D. G.; Beyer, J. E.; Carreon, R. M.; and Gehring, P. J. 1978. Long-term toxicological studies of 2,3,7,8-tetrachlorodibenzo-*p*-dioxin (TCDD) in laboratory animals. Paper read at Intern. Conf. on Health Effects of Halogenated Aromatic Hydrocarbons. 21–30 June at New York Academy of Sciences.

Kociba, R. J.; Keyes, D. G.; Beyer, J. E.; Carreon, R. M.; Wade, C. E.; Dittenber, D. A.; Kalnins, R. P.; Frauson, L. E.; Park, C. N.; Barnard, S. D.; Hummel, R. A.; and Humiston, C. G. 1978. Results of a two-year chronic toxicity and oncogenicity study of 2,3,7,8-tetrachlorodibenzo-*p*-dioxin (TCDD) in rats. *Toxicol. Appl. Pharmacol.* In press.

Langer, H. G.; Brady, T. P.; and Briggs, P. R. August 1973. The formation of dibenzodioxins and other condensation products from chlorinated phenols and derivatives. *Environ.*

Health Persp. Experimental Issue No. 5.

Leng, M. L. 1977. Comparative metabolism of phenoxy herbicides in animals. In *Fate of pesticides in large animals.* Eds., G. W. Ivie and H. W. Dorough, pp. 53–76. New York: Academic Press.

Levin, J. O.; Rappe, C.; and Nilsson, C. A. 1976. Use of chlorophenols as fungicides in sawmills. *Scand. J. Work Environ. and Health* 2:71–81.

Lucier, G. W., and McDaniel, O. S. 1978. Developmental toxicology of the halogenated aromatics: effects on enzyme development. Paper read at Intern. Conf. on Health Effects of Halogenated Aromatic Hydrocarbons. 21–30 June at New York Academy of Sciences.

Luster, M. I.; Faith, R. E.; and Clark, G. 1978. Laboratory studies on the immune effects of halogenated aromatics. Paper read at Intern. Conf. on Health Effects of Halogenated Aromatic Hydrocarbons. 21–30 June at New York Academy of Sciences.

McConnell, E. E., and McKinney, J. D. 1978. Exquisite toxicity in the guinea pig to structurally similar halogenated dioxins, furans, biphenyls and napthalenes. Paper read at 17th Ann. Meeting, Soc. of Toxicol., San Francisco.

McConnell, E. E., and Moore, J. A. 1978. Toxicopathology characteristics of the halogenated aromatics. Paper read at Intern. Conf. on Health Effects of Halogenated Aromatic Hydrocarbons. 21–30 June at New York Academy of Sciences.

McConnell, E. E.; Moore, J. A.; and Dalgard, D. W. 1978. Toxicity of 2,3,7,8-tetrachlorodibenzo-*p*-dioxin (TCDD) in rhesus monkey (*macaca mulatta*) following a single oral dose. *Toxicol. Appl. Pharmacol.* 43:175–87.

McConnell, E. E.; Moore, J. A.; Haseman, J. K.; and Harris, M. W. 1978. The comparative toxicity of chlorinated dibenzo-*p*-dioxins in mice and guinea pigs. Submitted for publication to *Toxicol. Appl. Pharmacol.*

Mahle, N. H.; Higgins, H. S.; and Getzendaner, M. E. 1977. Search for the presence of 2,3,7,8-tetrachlorodibenzo-*p*-

dioxin in bovine milk. *Bull. Environ. Contam. Toxicol.* 18:123.

May, G. 1973. Chloracne from the accidental production of tetrachlorodibenzodioxin. *Br. J. Indust. Med.* 30:276–83.

Mercier, M. J. 1976. 2,3,7,8-tetrachlorodibenzo-*p*-dioxin, an overview. Paper read at Meeting of Experts, 30 September–1 October, Giunta Regionale Lombardia, Milan, Italy.

Meselson, M. S.; O'Keefe, P.; and Baughman, R. The evaluation of possible health hazards from TCDD in the environment. Paper read at Symposium on the Use of Herbicides in Forestry, 21–22 February 1978, Arlington, Va.

Meselson, M. S.; Westing, A. H.; Constable, J. D.; and Cook, R. E. 2–9 March 1972. Preliminary report of Herbicide Assessment Commission of the American Association for the Advancement of Science. *Congressional Record.* 92nd Cong., 2d sess., 118, 6:6806–13.

Miller, R. A.; Norris, L. A.; and Hawkes, C. L. August 1973. The toxicity of 2,3,7,8-tetrachlorodibenzo-*p*-dioxin (TCDD) in aquatic organisms. *Environ. Health Persp.* Experimental Issue No. 5.

Milnes, M. H. 6 August 1971. Formation of 2,3,7,8-tetrachlorodibenzodioxin by thermal decomposition of sodium 2,4,5-trichlorophenate. *Nature* 232:395–396.

Moore, J. A.; Gupta, B. N.; Zinkl, J. G.; and Vos, J. G. 1973. Postnatal effects of maternal exposure to 2,3,7,8-tetrachlorodibenzo-*p*-dioxin (TCDD). *Environ. Health Persp.* 5:81–85.

Neal, R. A.; Beatty, P. W.; and Gasiewicz, T. A. 1978. Studies of the mechanisms of toxicity of 2,3,7,8-tetrachlorodibenzo-*p*-dioxin (TCDD). Paper read at Intern. Conf. on Health Effects of Halogenated Aromatic Hydrocarbons. 21–30 June at New York Academy of Sciences.

Neubert, D., and Dillmann, I. 1972. Embryotoxic effects in mice treated with 2,4,5-trichlorophenoxyacetic acid and 2,3,7,8-tetrachlorodibenzo-*p*-dioxin. *Naunyn-Schmiedebergs Arch. Pharmak.* 272:243–64.

Norback, D. H., and Allen, J. R. August 1973. Biological responses of the nonhuman primate, chicken, and rat to chlorinated dibenzo-*p*-dioxin ingestion. *Environ. Health Persp.* Experimental Issue No. 5.

Olie, K.; Vermeulen, P. L.; and Hutzinger, O. 1977. Chlorodibenzo-*p*-dioxins and chlorodibenzofurans are trace components of fly ash and flue gas of some municipal incinerators in the Netherlands. *Chemosphere* 6:455–59.

Oliver, R. M. 1975. Toxic effects of 2,3,7,8-tetrachlordibenzo-1,4-dioxin in laboratory workers. *Br. J. Indust. Med.* 32:49–53.

Pocchiari, F.; Silano, V.; and Zampieri, A. 1978. Human health effects from accidental release of TCDD at Seveso. Paper read at Intern. Conf. on Health Effects of Halogenated Aromatic Hydrocarbons. 21–30 June at New York Academy of Sciences.

Poland, A. P., and Glover, E. 1973. 2,3,7,8-tetrachlorodibenzo-*p*-dioxin: a potent inducer of δ-aminolevulinic acid synthetase. *Science* 179: 476–77.

Poland, A. P., and Glover, E. 1973. Studies on the mechanism of toxicity of the chlorinated dibenzo-*p*-dioxins. *Environ. Health Perspec.* 5:245-251.

Poland, A. P.; Glover, E.; Robinson, J. R.; and Nebert, D. W. 10 September 1974. Genetic expression of aryl hydrocarbon hydroxylase activity: induction of monooxygenase activities and cytochrome P_1-450 formation by 2,3,7,8-tetrachlorodibenzo-*p*-dioxin in mice genetically "nonresponsive" to other aromatic hydrocarbons. *J. Biol. Chem.* 249(17):5599–5606.

Poland, A. P., and Kende, A. 1976. 2,3,7,8-tetrachlorodibenzo-*p*-dioxin: environmental contaminant and molecular probe. *Fed. Proc.* 35:2404–11.

Poland, A. P.; Greenlee, W. F.; and Kende, A. S. 1978. Studies in the mechanism of action of the chlorinated dibenzo-*p*-dioxins and related compounds. Paper read at Intern. Conf. on Health Effects of Halogenated Aromatic Hydrocarbons, 21–30 June at New York Academy of Sciences.

Rappe, C.; Buser, H. R.; and Bosshardt, H.-P. 1978. Dioxins, dibenzofurans and other polyhalogenated aromatics—production, use, formation and destruction. Paper read at Intern. Conf. on Health Effects of Halogenated Aromatic Hydrocarbons. 21–30 June at New York Academy of Sciences.

———. 1978. Identification and quantification of polychlorinated dibenzo-*p*-dioxins (PCDDs) and dibenzofurans (PCDFs) in 2,4,5-T-ester formulations and herbicide orange. *Chemosphere.* In press.

Rappe, C.; Marklund, S.; Buser, H. R.; and Bosshardt, H.-P. 1978. Formation of polychlorinated dibenzo-*p*-dioxins (PCDDs) and dibenzofurans (PCDFs) by burning or heating chlorophenates. *Chemosphere* 7:269–81.

Reggiani, G. July-August 1977. Toxic effects of TCDD in man. Paper read at NATO Workshop on Ecotoxicology, at Guildford, England.

——. 1978. Medical problems raised by the TCDD contamination in Seveso, Italy. *Arch. Toxicol.* 40:161–88.

Rose, J. Q.; Ramsey, J. C.; Wentzler, T. H.; Hummel, R. A.; and Gehring. P. J. 1976. The fate of 2,3,7,8-tetrachlorodibenzo-*p*-dioxin following single and repeated oral doses to the rat. *Toxicol. Appl. Pharmacol.* 36:209–26.

Schulz, K. H. February 1968. Clinical picture and etiology of chloracne. *Arbeitsmedizin-Sozialmedizin-Arbeitshygiene* 3(2):25–29.

Schwetz, B. A.; Norris, J. M.; Sparschu, G. L.; Rowe, V. K.; Gehring, P. J.; Emerson, J. L.; and Gerbig, C. G. August 1973. Toxicology of chlorinated dibenzo-*p*-dioxins. *Environ. Health Persp.* Experimental Issue No. 5.

Shapley, D. 1973. Herbicides: agent orange stockpile may go to the South Americans. *Science* 180:43–45.

Shuman, R. M.; Leech, R. W.; and Alvord, E. C. 1975. Neurotoxicity of hexachlorophene in humans. *Arch. Neurol* 32(5): 320–25.

Sparschu, G. L.; Dunn, F. L.; and Rowe, V. K. 1971. Study of the teratogenicity of 2,3,7,8-tetrachlorodibenzo-*p*-dioxin in the rat. *Food. Cosmet. Toxicol.* 9:405–12.

Strik, J. J. T. W. A. 1978. Porphyrins in urine as indication for exposure to chlorinated hydrocarbons. Paper read at Intern. Conf. on Health Effects of Halogenated Hydrocarbons. 21–30 June at New York Academy of Sciences.

Susskind, R. R. 1955. A study of industrial acnegens. Paper

read at American Academy of Dermatology.

Telegina, K. A., and Bikbulatova, L. I. 1970. Affection of the follicular apparatus of the skin in workers occupied in production of butyl ether of 2,4,5-trichlorphenoxyacetic acid. *Vestn. Dermatol. Venerol.* (Moscow) 44:35–39.

Thiess, A. M. and Frentzel-Beyme, R. 1977. Mortality study of persons exposed to dioxin after an accident which occurred in the BASF on 13th November 1953. Paper read at the V. Medichem Congress, San Francisco.

Thigpen, J. E.; Faith, R. E.; McConnell, E. E.; and Moore, J. A. December 1975. Increased susceptibility to bacterial infection as a sequela of exposure to 2,3,7,8-tetrachlorodibenzo-*p*-dioxin. *Infection and Immunity* 12:1319–24.

Tung, T. T. 1973. Primary cancer of the liver in Vietnam. Translated from French. *Chirurgie* 99:427–36.

———. 1977. Pathologie humaine et animale de la dioxine. *Révue de Médecine* 14:653–57.

U.S., Air Force document. November, 1974. Final Environmental Statement. *Disposition of Orange Herbicide by Incineration.*

U.S., Congress, Senate, Committee on Commerce, Subcommittee on Energy, Natural Resources, and the Environment. 7 and 15 April 1970. *Effects of 2,4,5-T on Man and the Environment.* Serial 91–60.

U.S., Environmental Protection Agency. Advisory Committee on 2, 4, 5-T. 7 May 1971. *Report of the Advisory Committee on 2,4,5-T to the Administrator of the Environmental Protection Agency.*

U.S., Environmental Protection Agency. Dioxin Working Group. 2 April 1977. Dioxin: position document, draft.

U.S., Environmental Protection Agency. Office of Pesticide Programs. *Environmental Protection Agency Pesticide Programs. Rebuttable Presumption Against Registration and Continued Registration of Pesticide Products Containing 2, 4,5-T.* 43 *Federal Register* 17116–17118.

U.S., Environmental Protection Agency. Office of Pesticide Programs. Internal memorandum. 5 August 1975.

U.S., Executive Office of the President, Office of Science and

Technology. 1971. *Report on 2,4,5-T: A Report of the Panel on Herbicides of the President's Science Advisory Committee.*

Van Miller, J. P.; Lalich, J. J.; and Allen, J. R. 1977. Increased incidence of neoplasms in rats exposed to low levels of 2,3,7,8-tetrachlorodibenzo-*p*-dioxin. *Chemosphere* 6:537–44.

Van Miller, J. P.; Marlar, R. J.; and Allen, J. R. 1976. Tissue distribution and excretion of tritiated tetrachlorodibenzo-*p*-dioxin in non-human primates and rats. *Food Cosmet. Toxicol.* 14:31–34.

Verrett, 7 and 15 April 1970. Statement to Food and Drug Administration, HEW. In *Effects of 2,4,5-T on man and the environment,* pp. 190–203. Hearings before the Subcommittee on Energy, Natural Resources, and the Environment of the Committee on Commerce, U.S. Senate, 91st Congress, Second Session on Effects of 2,4,5-T on Man and the Environment. Washington, D.C.: U.S. Government Printing Office.

Vos, J. G. 1977. TCDD, effects/mechanisms. Background paper for the Royal Swedish Academy of Sciences Conference on chlorinated phenoxy acids and their dioxins: mode of action, health risks, and environmental effects, 6–10 February, at Stockholm.

Vos, J. G.; Kreeftenberg, J. G.; Engel, H. W. B.; Minderhoud, A.; and van Noorle Jansen, L. M. 1978. Studies on 2,3,7,8-TCDD induced immune suppression and decreased resistance to infection: endotoxin hypersensitivity, serum zinc concentrations and effect of thymosin treatment. Submitted to *Infection and Immunity.*

Vos, J. G.; Moore, J. A.; and Zinkl, J. G. August 1973. Effect of 2,3,7,8-tetrachlorodibenzo-*p*-dioxin on the immune system of laboratory animals. *Environ. Health Persp.* Experimental Issue No. 5.

Wallenfels, K. 1977. TCDD. Eine chemische und biochemische Analyse der Katastrophe von Ceveso. Commission for Civilian Population Protection, March-April.

Wassom, J. S., and Huff, J. E. The mutagenicity of chlorinated dibenzo-*p*-dioxins. Oak Ridge National Laboratory, Tennessee.

Whiteside, T. 1971. *The withering rain: America's herbicidal folly*. New York: E. P. Dutton.

———. *The New Yorker.*

7 February 1970. A reporter at large: defoliation. 32–69.

14 March 1970. Department of amplification. 124–29.

20 June 1970. Department of amplification. 78–95.

4 July 1970. Department of amplification. 64–70.

14 August 1971. On the use of herbicides in Vietnam. 54–59.

25 July 1978. A reporter at large: the pendulum and the toxic cloud. 30–55.

4 September 1978. A reporter at large: contaminated. 34–81.

Williams, D. T., and Blanchfield, B. J. November 1972. An improved screening method for chlorodibenzo-*p*-dioxins. *J. Assoc. Official Analytical Chemists* 55:1358–59.

Woolson, E. A.; Thomas, R. F.; and Ensor, P. D. J. 1972. Survey of polychlorodibenzo-*p*-dioxin content in selected pesticides. *J. Agr. Food Chem.* 20(2):351–54.

World Health Organization, International Agency for Research on Cancer. 1977. *Monographs on the evaluation of the carcinogenic risk of chemicals to man: some fumigants, the herbicides 2,4-D and 2,4,5-T, chlorinated dibenzodioxins and miscellaneous industrial chemicals.* Vol. 15. Lyon.

Young, A. L.; Thalken, C. E.; Arnold, E. L.; Cupello, J. M.; and Cockerham, L. G. 1976. Fate of 2,3,7,8-tetrachlorodibenzo-*p*-dioxin (TCDD) in the environment: summary and decontamination recommendations. Colorado: Department of Chemistry and Biological Sciences, USAF Academy.

Zelikov, A. Kh., and Danilov, L. N. 1974. Occupational dermatitis (acne) in workers engaged in the production of 2,4,5-trichlorophenol. Translated from Russian. *Sovetskaya Meditsina* 7:145–46.

Zinkl, J. G.; Vos, J. G.; Moore, J. A.; and Gupta, B. N. 1973. Hematologic and clinical chemistry effects of 2,3,7,8-tetrachlorodibenzo-*p*-dioxin in laboratory animals. *Environ. Health Persp.* 5:111–18.

Index